C0-ATK-168

The Limbic Brain

The Limbic Brain

Andrew Lautin, M.D.

Clinical Associate Professor of Psychiatry
Department of Psychiatry
New York University School of Medicine
New York, New York

and

Attending Psychiatrist
Department of Veterans Affairs Medical Center
Manhattan Campus
New York, New York

Kluwer Academic/Plenum Publishers
New York, Boston, Dordrecht, London, Moscow

Library of Congress Cataloging-in-Publication Data

Lautin, Andrew
The limbic brain/Andrew Lautin
p. cm. Includes bibliographical references and index.
ISBN 0-306-46086-6
1. 2.

ISBN 0-306-46086-6

233 Spring Street, New York, N.Y. 10013

http://www.wkap.nl/

10 9 8 7 6 5 4 3 2 1

A C.I.P. record for this book is available from the Library of Congress.

Printed in the United States of America

To

Arthur and Fredda

Terry, Sam, and Olivia

"*As Karl Pribram has said, 'Give me the freedom of seven neurons and I will take you from any one place to any other place in the nervous system.'*"

—Lacey

"*It is now known that the march of neural processing through the neocortex typically involves a sequence of association areas, and that a destination on the march seems invariably to be the hippocampus or the amygdala, or both.*"

—W. Nauta

We shall not cease from exploration
And the end of all our exploring
Will be to arrive where we started
And know the place for the first time.

T. S. Eliot, *Four Quartets*

ACKNOWLEDGMENTS

A man will turn over half a library to make one book
—Samuel Johnson

Acknowledgments can start anywhere. But one has to choose a place and time. Following Samuel Johnson's suggestion I will start with an expression of thanks to Erik Helman, and Tom Waugh, librarians at the Manhattan Veterans Administration Medical Center, New York, for their unceasing efforts in finding hard to acquire articles and books. From the department of Neurology at the New York University, I would like to express my gratitude to Drs. Edwin Kolodny, Herman Weinreb, and Richard Hanson, for providing the chance to present in lecture form the material that would subsequently become the chapters of this book. I am also indebted to Drs. Derby, and Cravioto for their open invitation to attend their distinctive, and engaging, weekly brain cutting lecture series, held in the basement pathology "theater" of the VA. Closest to home I would also like to thank Drs. Sam Gershon, John Rotrosen, Bert Angrist, and Adam Wolkin, from the department of psychiatry. Martin Roth (University of Cambridge psychiatrist) wrote that all psychiatric diagnoses are hypotheses. Sam Gershon's unit at Bellvieu followed this principal providing the open, and supportive environment that would allow my early interests in neuroanatomy to develop.

I also express my gratitude to Michael Hennelly and Jennifer Stevens at Plenum Publishers. Michael a senior editor at Plenum afforded a wisdom, and forbearance that enabled an inexperienced author to "almost" get the manuscript right. Ms. Stevens, served in an admirable capacity to help in the completion of this book.

CONTENTS

Chapter 2

CHAPTER SUMMARIES

The book, which is divided into five chapters, overviews basic concepts and principles in our understanding of limbic brain anatomy dating to Broca's proposal in 1878. In the first four chapters, the limbic brain is examined as a somewhat independent hemispheric domain. Broca's great limbic lobe, Papez's circuit, MacLean's limbic system, and Nauta's limbic midbrain form the conceptual core of chapters 1–4, respectively. The inextricable linkage of limbic and neighboring districts, implicitly acknowledged in the first four chapters, receives particular emphasis in chapter 5, a chapter, which addresses the conceptual issues put forth by Heimer and Wilson in their seminal publication on the ventral striatum.

CHAPTER 1

This chapter is intended to achieve four goals. An overview of several classical (19th century) descriptions related to the limbic brain concept. These include Rolando's *processo cristato*, Gerdy's *circonvolution annulaire*, and of course, Broca's great limbic lobe. A survey of the hippocampal formation and cortical limbic border in mid temporal section (at the body of the hippocampus) using recent illustrations, and terminology the chapter will provide an examination of how Broca's lobe relates to later conceptual developments in the fields of cytoarchitectonics, and comparative neuroanatomy will also be provided.

CHAPTER 2

In 1937, James Papez would see in the ensemble of the hypothalamus, cingulate gyrus, and their connections an anatomic theory of emotion. Francis Schiller remarked that Papez's idea would go on to create something of a stir. Kenneth Livingston referred to Papez's proposal as the most dramatic shift in emphasis in our view of limbic structures since the time of Broca. Seeming to capture the essential element in Papez's contribution, Paul Yakovlev wrote

> "In his now classic article of 1937, he [Papez] humbly proposed, and at some calculated risk, that 'emotions' have an anatomical 'mechanism,' and he marshaled evidence for the location of the mechanism in the brain. This empirical generalization contradicted the notion widely accepted in

> his time, that experience so exquisitely 'mental,' as are 'emotions are a preserve of psychology and that a generalization proposing an anatomical mechanism was a delusion."

This chapter examines the conceptual developments leading Papez to his proposal of an anatomic theory of emotion. Three conceptual underpinnings of Papez's idea are traced.

The results of transection experiments (Sherrington, Cannon, and Bard) conducted in the early 20th century which pointed to the importance of a forebrain mediator of emotional expression.

The emerging neuroanatomic, and neurobehavioral opinion in the first decades of this century, which maintained a primacy for medially arrayed hemispheric structures (i.e. the hippocampus, amygdala, and neighboring rostral diencephalic/hypothalamic loci) in the mediation of neurovisceral, endocrine, autonomic nervous system, and interoceptive processes. This view also held that the laterally displayed hemispheric wall structures served primarily somatic and exteroceptive processes.

The assumption (based on clinical, and theoretical grounds, and analogizing emotional awareness to other known sensory processes) that emotional consciousness required a cortical (not subcortical or peripheral) mediator. Papez would chose the mesocortex of the cingulate gyrus as the principal (but not the only) receptive area for the experience of emotion.

Cohering these areas, Papez put forth his idea of an anatomical circuit, which, in his own words, would serve as "a harmonious mechanism" mediating emotion. Papez's proposal was an early well principled anatomically determined theory of emotion. Emotional awareness/consciousness received *no* special privilege in Papez's theory, indeed the emotional sense would be brought into the same anatomic theater of explanation as the general somesthetic and special senses: i.e. similar to pain, temperature, and vision, emotional experience was accorded a thalamic nucleus and thalamocortical cortical relay. Coleridge's lines are evocative of Papez's grand conjecture.

> "I have not only completely extricated the notion of time, and space . . . but I trust that I am about to do more—namely, that I shall be able to evolve all the five senses, that is to deduce them from one sense & to state their growth, & the causes of their difference—& in the evolvement to solve the process of Life and Consciousness.
>
> SAMUEL COLERIDGE (1801).

Chapter two is devoted to a review of Papez's rather extraordinary rendering of core limbic function.

CHAPTER 3

From the early nineteenth century to Papez's proposal put forth in 1937, a favored view of higher mammalian limbic anatomy was that of a circumannular convolution, hugging the mid sagittal plane, and limned (bordered) by the limbic fissure. However within two decades of Papez's paper the limbic brain concept would be transformed to include a consortium of neural structures extending rostral to the limbic fissure, yet intimately connected with Papez's core circuit elements. Brought into the limbic ambit were the orbitofrontal cortex, also perihippocampal districts. No longer constrained to the midsagittal plain the rather exuberantly conceived limbic coalition would be called a *limbic system*. One of the principal figures who brought about this transformation was Paul MacLean. MacLean also engaged in an enterprising endeavor to place the limbic system of mammals in a broadly conceived evolutionary context. In this chapter, we examine the contribution of MacLean to limbic brain neuroanatomy and function.

Chapter 3 also includes an appendix addressing the work on limbic brain anatomy of Paul Yakovlev (1894–1983). Although Yakovlev's language and approach has been characterized as unusual—indeed Herrick, commenting on Yakovlev's seminal publication on the limbic brain said he agreed with its argument but "would have put it differently"—Yakovlev's paper on limbic brain anatomy stands alongside MacLean's and Papez, it is one paper from this famous limbic trilogy. We hope the appendix to this chapter conveys a sense of Yakovlev's contribution to the concept of the limbic brain.

CHAPTER 4

If a general trend can be said to characterize neuroanatomic research on the limbic brain at mid century then surely it was the extension of anatomic borders of limbic system to even further rostral and caudal domains of the neuraxis. Spearheaded by Walle Nauta, the limbic consortium would expand further to embrace more firmly both neocortical and midbrain structures. In this chapter we explore Nauta's concept of the limbic midbrain and examine his views on prefrontal to basal hemispheric limbic connections.

CHAPTER 5

This chapter is centered on an examination of the publication of Heimer and Wilson. This paper helped to better establish the nature of the anatomical interface of limbic and striatal domains. Referred to by McGeer as the twin galaxies of the inner universe, a discussion of the linkage of striatal and limbic anatomic districts also serves to de-emphasize the concept of an independent limbic brain domain, and ends our introduction to limbic anatomy. Interestingly, Heimer and Wilson, make the point that a strict interpretation of then extant limbic anatomical canons would act to forestall broad acceptance of limbic striatal integration in the first half of this century. In this chapter, avoiding this potential pitfall, we employ the basic principles of limbic brain anatomy reviewed in the earlier chapters to better appreciate the nature of Heimer and Wilson's findings. Chapter 5 concludes with a brief examination of limbic striatal integration, as put forth by Gordon Mogenson in his oft-referenced paper on limbic motor integration.

REFERENCES

1) Broca, P. (1878). Anatomie comparee des circonvolutions cerebrales. Le grand lobe limbique et la scissure limbique dans la serie des mammiferes, *Rev. Anthropol. 1*, Ser. 2, 385–498.
2) Papez, J. W. (1937). A proposed mechanism of emotion, *Arch. Neurol. Psychiatry*, 38, 725–743.
3) Yakovlev, P. (1972). A proposed definition of the limbic system; in *Limbic System Mechanisms and Autonomic Function* (C. H. Hockman ed.), Springfield Illinois, 241–283.
4) MacLean, P. (1952). Some psychiatric implications of physiological studies on frontotemporal portion of limbic system (visceral brain), Electroencephalogr, *Clin. Neurophysiol.* 4, 407–418.
5) Yakovlev, P. (1978). Recollections of James Papez and comments on the evolution of the limbic system concept (As told to Ken Livingston), in *Limbic Mechanisms* (K. E. Livingston and Oleh Hornykiewicz eds.), Plenum Press, 351–354.
6) Yakovlev, P. I. (1948). Motility, behavior, and the brain. Stereodynamic organization and neural coordinates of behavior, *J. Nerv. Ment Dis*. 107, pp. 313–335.
7) Nauta, W. (1958). Hippocampal projections and related neural pathways to the mid-brain in the cat, *Brain* 81, 319–340.
8) Heimer, L., and Wilson, R. D. (1975). The subcortical projections of the allocortex Similarities in the neural associations of the hippocampus, the piriform cortex, and the neocortex, in: *Proceedings of the Golgi Centennial Symposium* (M. Santini, ed.), Raven Press, New York, 177–193.
9) Mogenson, G. J., Jones, D. L., and Yim, C. Y. (1980). From motivation to action: Functional interface between the limbic system and the motor system, *Progress in Neurobiology*, pp. 69–97.

INTRODUCTION

Nearly, 50 years ago, Karl Pribram in a discussion section accompanying MacLean's proposal of a limbic system, criticized the visceral or limbic brain concept as theoretically too vague and cumbersome. In a recent review of the limbic system, Swanson points to Brodal's criticism that the discovery of connections of limbic structures with virtually all parts of the nervous system render the concept of the limbic system useless, and better abandoned. Additional dissatisfaction surrounding the limbic brain concept stems from the feeling that it is historically inert (an antiquated 19th century construct). In our current age of neural networks, and parallel distributed process it is of little value, merely an historical curio. So why then this introduction to limbic brain anatomy? We offer several interrelated rationales behind our labors.

Recapitulation in the Service of Education: Although concepts had evolved in the second half of this century which effectively overthrew the idea of relatively isolated hemispheric districts (i.e. striatal, cortical, and limbic), parsing the hemisphere into these three districts was an important preliminary step achieved by our forebears in their efforts to understand the large scale structure of the higher mammalian cerebral hemisphere. An examination of how the limbic brain concept came to be provides an opportunity to recapitulate the process of exploration, discovery, and understanding as it relates to one of these principle hemispheric domains.

The Strange, Natural Beauty of this Neural Landscape: The neural terrain surrounding the epithelial margin of the hemisphere (Broca referred to this cortical limbic area as the threshold of the hemisphere) has engaged the interest of anatomists for the last two centuries. The following comments by four of the leading anatomists of the 20th century underscore the attraction, and importance of this area:

> "... I may say that one of the stimuli which led me to scrutinize the hippocampus and the dentate fascia was the elegant architecture shown by the cells and layers of these centres as revealed by the illustrious Golgi in his great work. In fact, the hippocampus and the dentate fascia are adorned by many features of the pure beauty of the cerebellar cortex. Their pyramidal cells, like the plants in a garden—as it were, a series of hyacinths—are lined up in hedges which describe graceful curves."
>
> RAMON CAJAL (4, p. 415)

> Certain of the most excellent contributions in to the history of comparative neurology have dealt with the hippocampal region of mammals, and this is true whether the general relations of the parts in the different mammals are taken into consideration . . . or the finer details of structure are considered . . .
>
> ARIENS KAPPERS, HUBER, CROSBY, 1936 (5, p. 1410)

> Little wonder, all in all, that Broca decided he had found the edge of the cerebral cortex, at least in the temporal lobe, and little wonder that he initially chose to call his discovery the great lobe of the hem. Here in the temporal lobe, the edge is clearly arrayed in a way suggesting the strategy of a seamstress who finishes off the hem of a garment by folding its material several times and even placing over its end an extra piece of fabric before stitching it all together.
>
> NAUTA, 1986 (6, pp. 274–275)

> Therefore, as I guide the reader through the strange landscape of the hippocampal region, I remind him of the history of search and discovery here, a history that can be perused with pleasure and scholarly satisfaction." (White, Yakovlev) . . .
>
> J. ANGEVINE, 1978

A principle goal of this book is to follow in the footsteps of such intrepid explorers, journey into this alluring neural landscape; and learn the secrets of its design, and function.

The Anatomy of Consciousness—Early Theories: Throughout much of the last century, more attention was directed to the limbic neural consortium (including the rostral brain stem) in the mediation of consciousness and emotional awareness, than on structures of seemingly equal complexity, such as the cerebellum. Why is this? Several anatomic theories which have attributed a central role of limbic (or equivalently centrencephalic) anatomy to the mediation of consciousness have been put forth, a partial list includes Herrick's view (and it would seem Broca's view as well for Broca also spoke of this issue), which held that the olfactory system established motivational contexts, and directly motivated behaviors. Papez's proposed circuit elaborating emotion, MacLean's limbic system, Yakovlev's paper on the three spheres of motility and behavior, Olds and Milner's discovery of the reward systems of the brain, notably their indwelling electrodes were positioned in what they referred to as the rhinencephalon, high rates of self stimulation were recorded in the mammillothalamic tract and cingulate cortex (11, p. 425). These two structures are cardinal elements in Papez's circuit. Heimer and Wilson's proposal (actually a re-conceptualization) of the basal forebrain and associated limbic loci as a macro-structure enabling motivation to become action, and relatedly Mogenson's celebrated paper addressing the same theme of motivation transmuted into action in the midbrain and basal forehair. An historical overview of our attempts to relate brain, consciousness and emotionality, would be incomplete without surveying such limbic centered proposals.

In Search of the Unity of Type: Broca' s proposal of the great limbic lobe (a great inner ring of tissue embedded in all mammalian brains) can (and should) be seen as an elegant contribution by anatomists to reduce neuroanatomic (cerebral hemispheric) diversity to one or very few underlying architectural themes or building blocks. This grand romantic vision (the aspiration of transcendental morphology) had received its clearest voice earlier in 19th century France with Etienne Geoffroy Saint-Hilaire's brazen assertion that, "*There is philosophically speaking only a single animal.*" Broca's fellow countrymen, Achille Foville, with his concentric ring model of the hemisphere; and Pierre Gerdy's annular circonvolution, look backward to Geoffroy, and presage Broca's great limbic lobe proposal. In the examination of limbic lobe description by 19th century anatomists, one is presented with some of the early and more alluring large scale geometric models put forth to explain and unify mammalian hemisphere architecture.

The Development of Our Conceptual Understanding in Neuroanatomy: Placing Broca's Great Limbic Lobe in a Continuum of Ideas Dating From the Early Nineteenth Century to the Present Day:

Examining both the early 19th century origins of Broca's proposal on the limbic brain, and the relationship of Broca's construct to ideas on hemispheric anatomy which came after his proposal is an important exercise which, if followed, can help shed light on the development of our concepts of hemispheric neuroanatomy. The following brief examples trace developments in limbic lobe anatomy after Broca demonstrated the fruitfulness of this exercise. Broca put forth his proposal twenty years before the advent of "modern" cytoarchitectonic description (remember that Broca was working at a relatively early age in the science of microscopy). With the maturation of cytoarchitectonics in the closing decades of the 19th century, it would be seen that the hemispheres evolutionary old cortex is contained in Broca's idea of the limbic lobe (or equivalently Broca's lobe would be seen as compromising the hemisphere allo-and periallocortical investments (9, p. 259)). The first characterization of Broca's lobe employing recognizable modern cytoarchitectonic lexicon may date to E. A Schaefer's Text-Book of Physiology (1900); More recent ideas in the field of comparative hemispheric topology can also be employed to recast Broca's proposal. Incorporating these ideas with their attendant terminology reveal Broca's lobe to be a derivative of the medial pallium and neighboring sectors of the dorsal and lateral pallium.

The Uniqueness of the Temporal Lobe and Limbic System: A point made most tellingly in Gloor's recent text, is that the temporal lobe and its limbic components particularly in higher primates represents a macrostructure for a privileged interaction of isocortical, allocortical, and subcortical (amygdala) neural structures. Such an anatomic interrelationship is not seen to the same degree in less temporalized forms. The conjunction of visual input (inferior temporal lobe) and auditory input (superior temporal lobe) at the hilar limbic threshold compels one to reflect on the behavioral implications of this unique anatomy. Gloor comments: "*This evolutionary development probably has profound implications for the way we as humans experience and interact with the world around us*." (15, p. 12). An examination of the temporal lobe and its limbic infrastructure would appear motivated in part by the very structure itself.

REFERENCES

1) MacLean, P. (1952). Some psychiatric implications of physiological studies on frontotemporal portion of limbic system (visceral brain), Electroencephalography: The basal and temporal regions, *Yale J. Biol. Med.* 22, 407–418.
2) Swanson, L. W. (1987). Chapter Entitled, Limbic System, in *Encyclopedia of Neuroscience* (G. Adelman, ed.), Birkhauser.
3) Broca, P. (1878). Anatomie comparee des circonvolutions cerebrales. Le grand lobe limbique et la scissure limbique dans la serie des mammiferes, *Rev. Anthropol.* 1, Ser, 2, 385–498.
4) Ramon y Cajal, S. (1996). *Recollections of My Life*. MIT Press.
5) Ariens Kappers, C. U., Huber, G. C., and Crosby, E. C. (1936/1967). *The Comparative Anatomy of the Nervous System of Vertebrates, Including Man*. Hafner Publishing Company, New York.
6) Nauta, W. J. H., and Fiertag, M. (1986). *Fundamental Neuroanatomy*, W. H. Freeman & Company, New York.
7) Herrick, C. J. (1933). The functions of the olfactory parts of the cerebral cortex, *Proceeding of the National Academy of Sciences*, pp. 7–14.
8) Papez, J. (1937). A proposed mechanism of emotion, *Arch. Neurol. Psychiatry*, 38, 725–743.
9) MacLean, P. (1990). *The Triune Brain in Evolution: Role in Paleocerebral Functions*, Plenum Publishing.
10) Yakovlev, P. I. (1948). Motility, behavior, and the brain. Stereodynamic organization and neural coordinates of behavior, *J. Nerv. Ment Dis.* 107, pp. 313–335.
11) Olds, J., and Milner, P. (1954). Positive reinforcement produced by electrical stimulation of septal areas and other regions of the rat brain. *J. Comp. Physiol. Psychol.* 47, 419–427.

12) Heimer, L., and Wilson, R. D. (1975). The Subcortical Projections of the Allocortex: Similarities in the Neural Associations of the Hippocampus, the Piriform Cortex, and the Neocortex, in: *Golgi Centennial Symposium. Proceedings* (M. Santini, ed.), Raven Press, New York, pp. 177–193.
13) Mogenson, G. J., Jones, D. L., and Yim, C. Y. (1980). From motivation to action: Functional interface between the limbic system and the motor system, *Progress in Neurobiology*, pp. 69–97.
14) Northcutt, R. G. (1969). Discussion of the preceding paper. Annals New York Academy Sciences, 167: 180–185.
15) Gloor, P. (1997). *The Temporal Lobe and Limbic System*, Oxford University Press.

CHAPTER ONE

BROCA'S LOBE

"... il constitue dans le manteau une division primaire, une division fondamentale qui est plus qu'un lobe, qui renferme d'ailleurs plusieurs lobes, et que le simple nom de lobe ne caracteriserait pas suffisamment: ***je l'appellerai done le grand lobe limbique.****"*

1, p. 392 BROCA's emphasis

"Le grand lobe limbique existe, plus ou moins distinct, plus ou moins volumineux et plus ou moins complet, chez tous les mammiferes."

1, p. 398

HEMISPHERIC DOMAINS: DISTINCTIVE FEATURES

There are several approaches to learning basic principles of mammalian hemispheric anatomy. One approach is to conceive of the mammalian hemisphere as constituted by three great domains: the neocortex, striatum, and limbic lobe. Such an approach is well motivated for two reasons. It recapitulates, in part, the development of our concepts of the higher mammalian hemisphere (indeed, to paraphrase Nauta (2) during the middle decades of this century conceptual developments in the evolution of our ideas regarding hemispheric architecture initially argued for these three domains to be (largely) independent primary hemispheric components). In the higher mammalian hemisphere each of these divisions can be said to exhibit its own distinctive anatomic features: and gaining an appreciation of these features provides a foundation for conceptualizing hemispheric architecture.

For example, the neocortex is distinguished as the outermost and largest division of the hemisphere; it is additionally characterized by distinct laminations, see figure 1.1A. The corpus striatum the massive core of

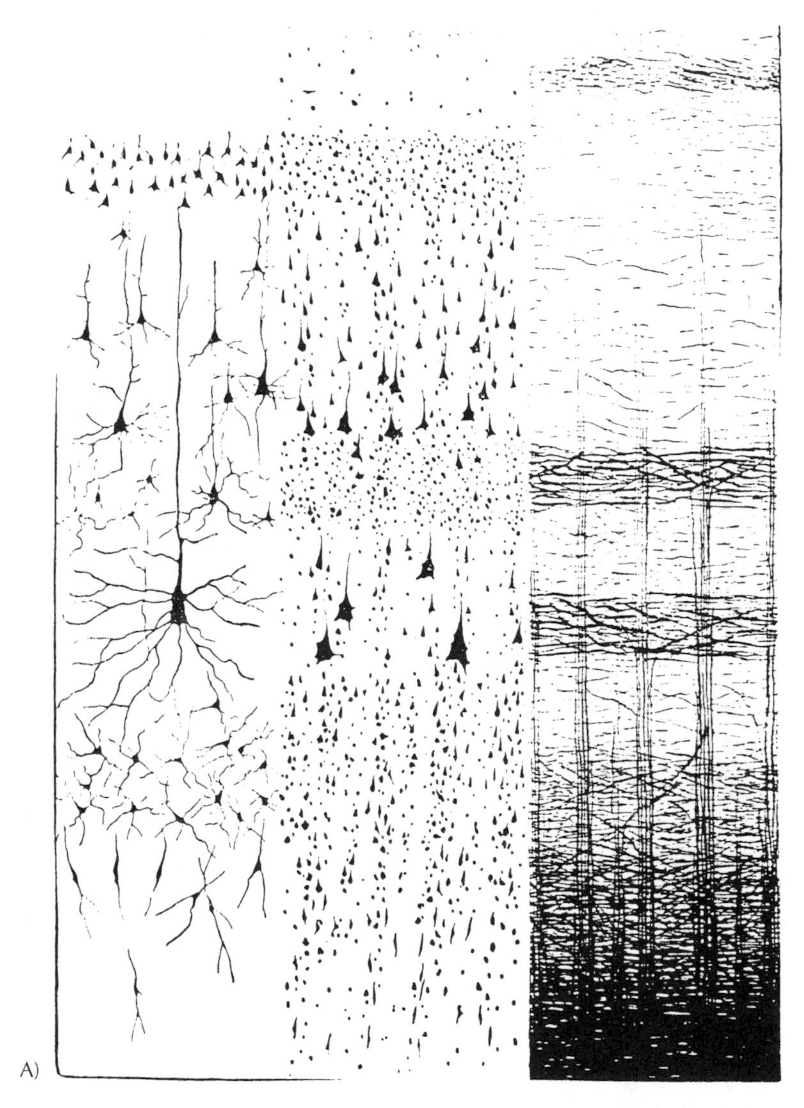

Figure 1.1. A) The six layers of the neocortex are evident in silver, Nissl, and myelin stain. B) Brain stem of a sheep, cerebral hemispheres removed. Thomas Willis's text, illustration by Christopher Wren (Cerebri Anatomie, 1664). The initial use of the term striatum dates to Willis. Here we see "in the foremost part of the brainstem prominent, lentiform bodies . . . which are characterized by up-and-downward radiating medullated fibers producing a striped appearance on their cut-surface" "Corpus striatum sive medullae oblongatae apices sunt duo prominentiae lentiformes." C) Sagittal section of the brain of a green anolis lizard demonstrating the defining feature of striatal anatomy (the reptilian representation of Wilson's pencils). Reference to the lizard's brain emphasizes the great phylogenetic age of the striatal core—the striatum is a common element in the brains of birds, reptiles, and mammals. Figure 1A, p. 227 from (3) Figure 1B and Willis citation from (4). Figure 1C (5, p. 37).

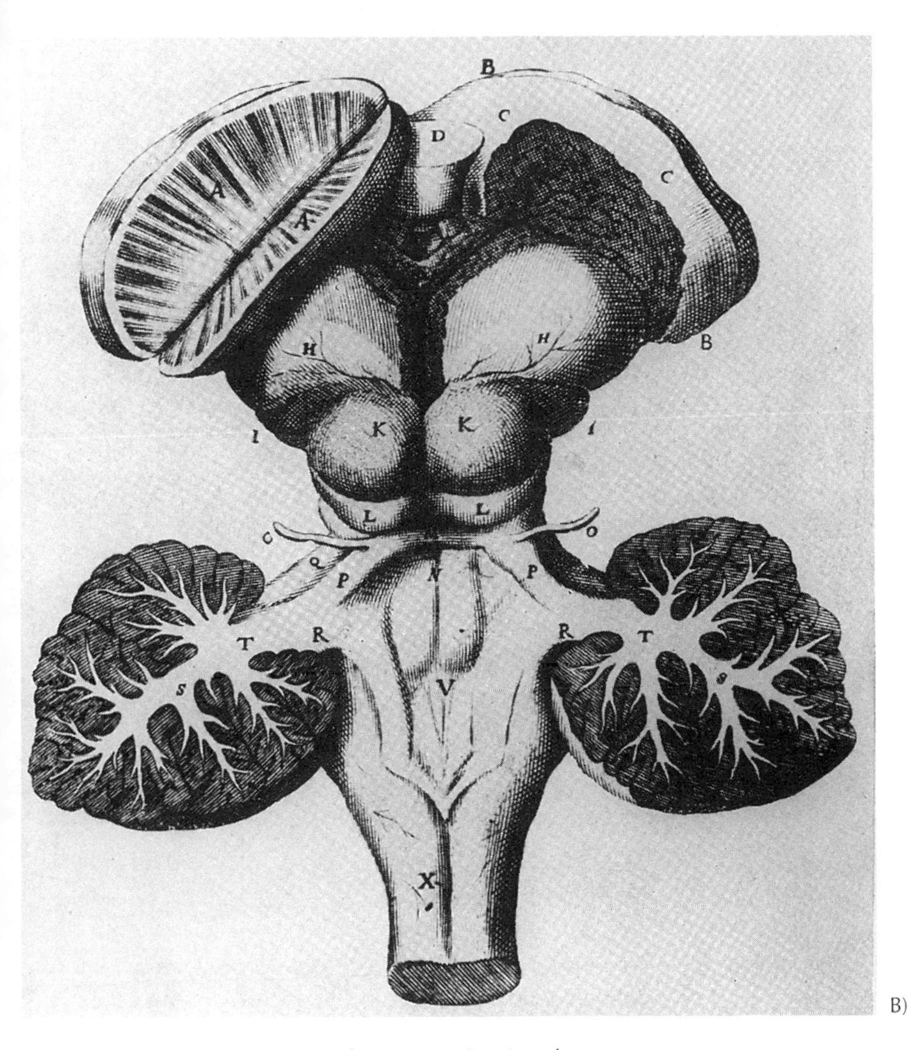

Figure 1.1. *Continued.*

the hemisphere composed principally of the caudate and putamen is recognized by the presence of distinct striations visible on gross inspection. Wilson's pencils,[1] more commonly referred to as radial fibers, in their

[1] Making lesions in the putamen, Samuel Kinnier Wilson (1878–1937) demonstrated that the fibers from the striatal cells in the monkey collect in bundles leading to the globus pallidus. Wilson referred to these bundles as pencils, hence the term Wilson's pencils. These

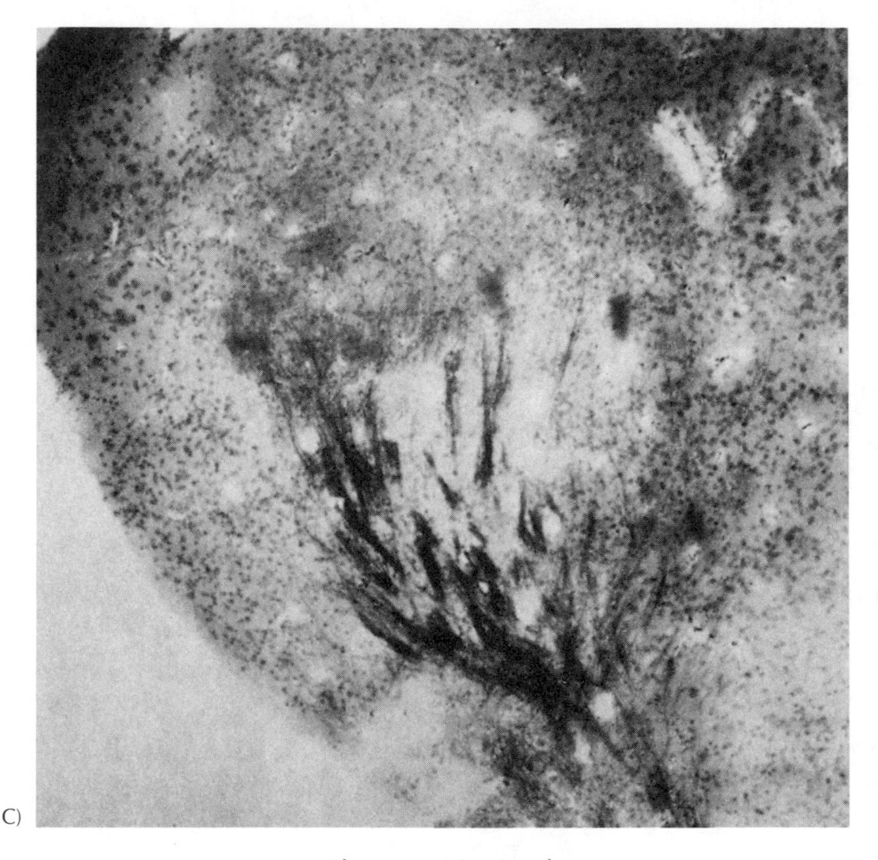
C)

Figure 1.1. *Continued.*

passage through the gray parenchyma of the lenticular nucleus are responsible for the distinguishing attribute of striatal anatomy, see figure 1.1B & 1.1C. (The term striatal has a more general meaning, i.e., the term is simply a descriptive term indicating regions consisting of large masses of gray matter with bundles of myelinated axons passing through them, thus imparting a striated appearance. The term is applied very broadly in regard to the avian telencephalon (6). The limbic district includes (and is ultimately defined by the fact that it so includes) the edge of the cortex. Broca

---

myelinated fibers converging on the globus pallidus mark out the principle neural traffic flow in the lenticular nucleus, a pattern that resembles a radial spoke wheel pattern converging on the globus.

emphasized several features of the limbic district, which still inform present day description. An examination of Broca's contribution follows.

LIMBIC LOBE: DISTINCTIVE FEATURES—BROCA'S PROPOSAL

In 1878, Paul Broca (1824–1880) authored his seminal publication on what he called the great limbic lobe. Broca drew attention to three basic features of mammalian limbic brain anatomy, see Table 1.

In the following two citations Broca draws our attention to the cortical limbus (cortical epithelial margin) dubs the larger convolution that embraces the cortical limbus as the limbic convolution, and introduces the term limbic fissure: these are the distinctive features listed in the first table.

> "The cerebral mantle covers all of the hemisphere with the exception of a very circumscribed region seen on the internal surface . . . While that portion of the internal surface which is left uncovered by the cerebral mantle does not represent an actual opening, it is, in a way, the entrance and exit of the hemisphere, a feature which in Latin terminology. . . . may be rendered by the word 'limen'. As it surrounds the threshold of the hemisphere the mantle forms a border which resembles the circular edge of a purse. **Hence I am calling this border the limbus of the hemisphere, and the convolution that forms it the 'limbic convolution'.**"
>
> 1) P. 385, 1878, Trans. From 2nd, sentence onward, SCHILLER (7) p. 256, emphasis added.

> "The great lobe, completely covers as one continuous unit the entrance of the hemisphere, however one distinguishes three component lobes.

Table 1.1. Basic Features of Limbic Anatomy as Put Forth by Broca

Cortical Limbus (Grey—White Border)	The cortical mantle (cerebral cortex) *DOES NOT* cover the entirety of the hemispheric wall but terminates at a rather sharply defined border (the cortical limbus).
Limbic Convolution	Broca's limbic convolution contains the cortical limbus. It was Broca's special contribution to draw attention to the constancy and robustness of this convolution (gyrus fornicatus) in a large series of mammals.
Limbic Fissure	The limbic convolution is circumscribed throughout its extent by the cortical epithetial border on one side, and the limbic fissure on the other.

> They are: 1^0 above, the lobe of the corpus calleux [cingulate gyrus], which forms the superior arc . . . ; 2^0 below, the hippocampal lobe, which form the inferior arc; 3^0 in front, the olfactory lobe which forms the base and in the manner of a racket handle forms the anterior aspect of the lobe. The anterior part of the olfactory lobe is free below the anterior extremity of the hemisphere, **the rest of the great lobe is circumscribed by a fissure which separates it and the cerebral convolutions and I call it the limbic fissure.**"
>
> (1) pp. 399–400, emphasis added.

DISTINCTIVE FEATURES OF LIMBIC LOBE ANATOMY: A CLOSER EXAMINATION

In the following two subsections we will expand our analysis of the basic elements of Broca's limbic lobe. Our focus in this and the following

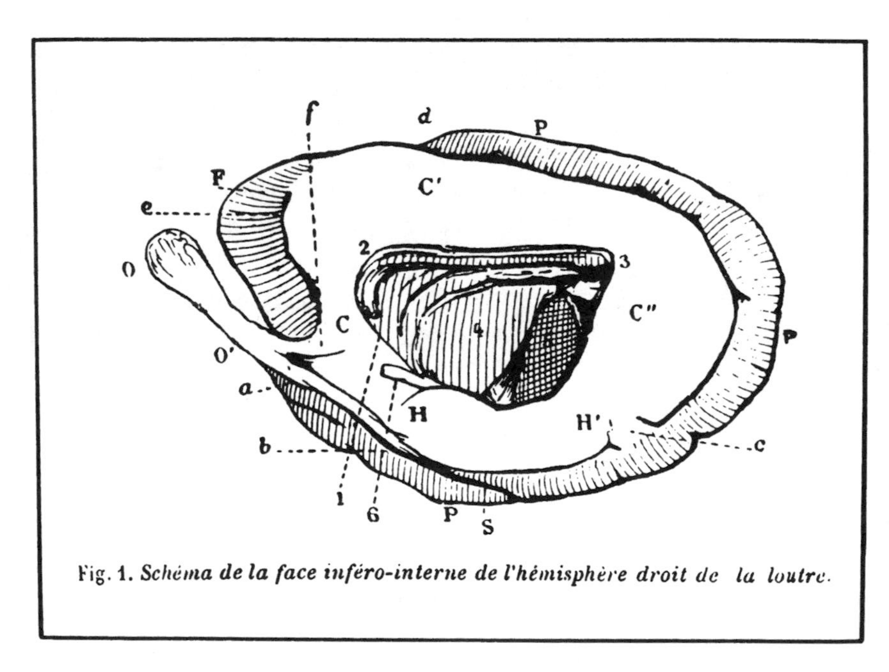

Figure 1.2. The medial and inferior aspect of the otter's brain appears as the first drawing in Broca's paper. H-H' refers to the hippocampal gyrus; CC'C' refers to the corps calleux or the cingulate gyrus; O, identifies the olfactory bulb, and surrounding olfactory lobe; 1-2-3, the corpus callosum, 4, the thalamus. Broca's findings were based on a comparative study of a series of mammals, in the 36 figures in his article, he illustrated different views of the lobe in 17 mammalian species.

sections will be oriented to the human brain (highly temporalized, robust development of the gyrus fornicatus).

Limbic Fissure

The limbic fissure can be envisioned as mapping out an arc, or part of a spiral, see figure 1.3. Notably the fissure is a variable and inconsistent finding in lower mammals; it is also often discontinuous in higher mammals. Following MacLean (5, p. 259) and Elliot Smith (8, p. 312) several principal parts of the limbic fissure are recognized. 1) At the rostrofrontal aspect is found the genual (L. *genu*, knee) sulcus which begins its course below the anterior edge of the corpus callosum, and curves upward, and around while remaining parallel to the genu of the corpus callosum. 2) Posteriorly the fissure is formed by the splenial sulcus which curves around and remains parallel to the splenium (L. *splenium*, bandage, refers to a widening or spreading of the corpus callosum in its posterior aspect). Elliot Smith concluded that the genual and splenial sulci were the most constant features of the limbic fissure. 3) For the inconstant sulcus intercalated between these two, Elliot Smith used the term intercalary sulcus. The intercalary sulcus

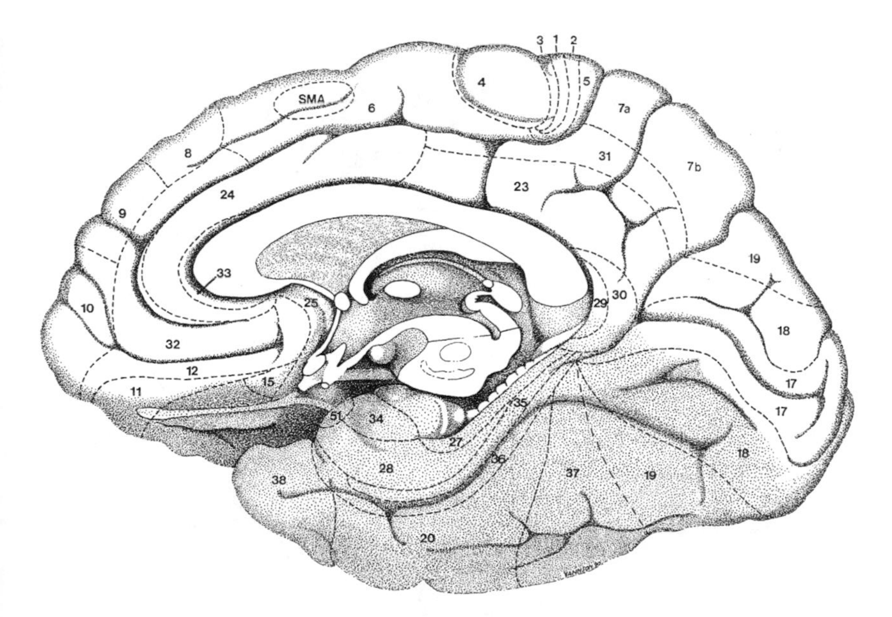

Figure 1.3. Midsagittal section through the hemisphere in the adult, demonstrating the limbic fissure and convolution. The brainstem has been removed to better illustrate the gyrus fornicatus throughout its course. The divisions of the hypothetical limbic fissure are discussed in the text. Figure from Duvernoy (10).

runs above and parallel to the corpus callosum. In primates the intercalary and genual sulcus are linked as the cingulate sulcus (aka. callosalmarginal sulcus, this aspect of the limbic fissure forms the *outer* boundary of the cingulate gyrus). 4) In its posterior aspect, the recurved splenial sulcus is continued by the anterior calcarine sulcus (i.e. the anterior aspect of the parieto-occipital sulcus). 5) The limbic fissure is continued by the collateral sulcus; however, linear continuity of the anterior calcarine sulcus with the collateral sulcus is never encountered. The collateral sulcus is located at the tentorial surface of the hemisphere. 6) The collateral sulcus in turn is continued by the final component, the rhinal fissure. Continuity of the collateral sulcus with the rhinal sulcus is commonly but not always encountered, see discussion by Gloor (9, p. 329).

Limbic Convolution

The annular ring of tissue on the medial face of the cerebral hemisphere enclosed by the limbic fissure consists of the subcallosal area, and the gyrus fornicatus. The fornicate gyrus includes, in its superior aspect, the

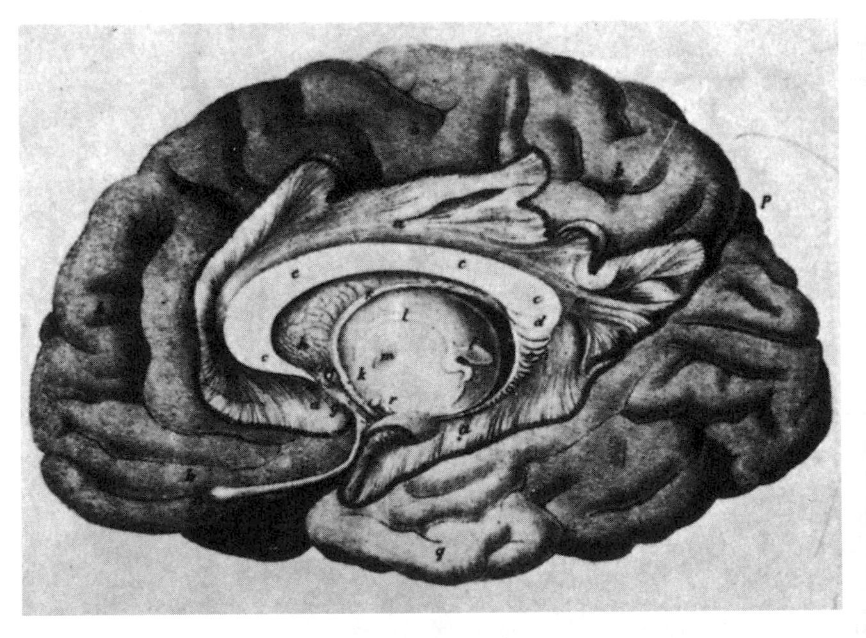

Figure 1.4. Luigi Rolando (1773–1831) dissected this area nearly forty years before Broca; he referred to this circumannular convolution as the *processo cristato*. Figure from (12).

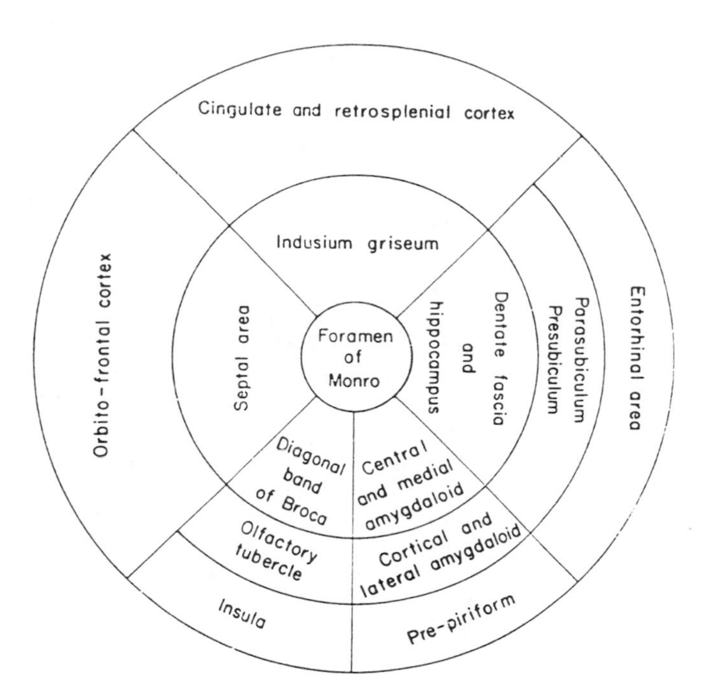

Figure 1.5. This view of the limbic lobe is achieved by collapsing three-dimensional hemispheric anatomy into two dimensions and removing the confound of the corpus callosum. The circular nature of the limbic area and the concentric ring appearance of the hemisphere are graphically realized in this idealized schematic. White emphasized the similarity of his model to Theodor Meynert's description. The concentric ring perspective on hemispheric form (with the limbic components constituting the inner rings) has a rich tradition dating at least to Foville in the nineteenth century, figure from White, with permission (13, p. 21).

cingulate, and in its inferior aspect, the parahippocampal gyrus. A clock analogy helps identify the major elements of the limbic convolution (5). Located at nine o'clock is the olfactory bulb (*en avant le lobe olfactif dont la base*). Proceeding clockwise, the paraseptal area is first encountered, with subsequent passage along the along the cingulate gyrus. At three o'clock, the parasplenial aspect of the cingulate is located. Cingulate parahippocampal confluence is encountered between three and 6 o'clock. Broca likened his great limbic lobe to a racket with the handle formed by the olfactory bulb and peduncle, and the racket head by the limbic (fornicate) convolution (7, p. 258).

Broca's limbic convolution did not stop at the parahippocampal aspect of the fornicate gyrus but extended along the parahippocampal

Table 1.2. Early Descriptions Prior to Broca

Johann Reil (1809)	*Windung der inneren Flache der Hemisphere (Langenwindung)*
Luigi Rolando (1830)	*Process Cristato*—Rolando described the processo cristato, which included the medial olfactory root, the cingulate and the hippocampal gyrus, see figure 1.4. Rolando's account was similar to Reil's.
Pierre Gerdy (1838)	*La Circonvolution Annulaire*—Gerdy's annular circonvolution represented a marginal band on the medial hemispheric wall surrounding the hemispheric hilum. The convolution included the gyrus fornicatus, which he described as being connected anteriorly with the olfactory lobe, and posteriorly with the hippocampi. He regarded his ring as being interrupted by the Sylvian fissure (11, p. 157).
Achille Foville (1839)	*La Circonvolution de L'ourlet*—a pull string purse or a hem: A popular concept among many 19th century anatomists was to view the hemisphere as constructed of generous loops (concentric rings). A. Louis Foville perceived three concentric loops. The innermost he called the circonvolution de l'ourlet. In his use of the term grand/great limbic lobe, Broca shared Foville's lack of concern for the restrictions of conventional lobar anatomy.
Achille Foville, & Pierre Gratiolet	*La Grande Overt de L'Hemisphere*—Refers to the opening into the inner recess of the hemispheric wall, i.e., the lateral ventricle. The immediately bordering (limbic) structures surrounding the opening are the choroidal fissure, white matter (fimbria) and gray matter (dentate gyrus).
Theodor Meynert (1872)	Meynert viewed the cerebrum as composed of an inner and outer ring of tissue centered on the foramen of Monro. Meynert's inner ring conforms to Broca's limbic lobe, see figure 1.5.

As emphasized by MacLean, also Schiller, although Broca coined the term limbic lobe and endowed it with special meaning he did not claim priority in reporting on this circumannular convolution. Even a cursory examination of the table demonstrates that the inner circular ring of nervous tissue, which constitutes Broca's limbic lobe, was a well-recognized feature of the mammalian brain early in the 19th century. Gustave Schwalbe's concept of the falciform lobe (L. *falx*, sickle), similar to Broca's lobe was put forth in 1881, three years after Broca's proposal. Although not mentioned in the table, Meyer, in his extensive review of the historical literature credits Sylvius (1663), also Willis (1664) as the first to refer to a limbus or border region (19, p. 111).

dentate continuum as far as the cortical limbus, best appreciated in coronal section, see figures 1.6, 1.7, 1.8, 1.9, & 1.10. With the inclusion of the cortical limbus, Broca was able to make his proclamation that his great limbic lobe which encompassed the cortical border provided, in a way, (as he put it) the entrance and exit to the hemisphere, see (p. 5).

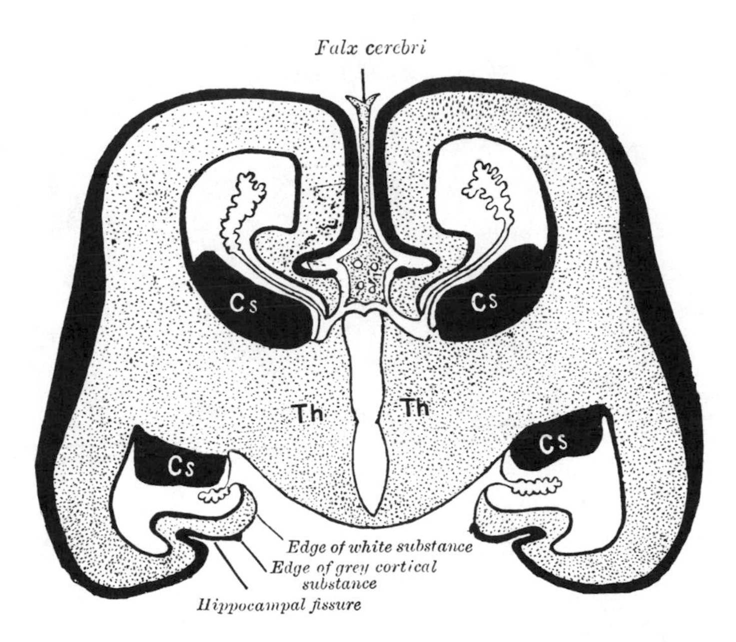

Figure 1.6. In this diagram, His, has presented a simplified view (absent the confounds of cortical laminations, and relatedly the corpus callosum) of the medial hemispheric wall and limbic boundary. Note that the gray white border (cortical limbus) is seen twice in cross section, in its inferior and superior aspect: in lower mammals, curvature of the hemisphere is ***NOT*** advanced enough and the limbus would be cut just ***ONCE***. Figure from Gray's Anatomy, with permission (15).

Broca's construct serves to direct our focus to the limbic lobe and cortical border. In the following sections we will pursue the cortical limbic terrain to an extent not possible in Broca's time.

WHY DID BROCA USE THE APPELLATION: THE GREAT LIMBIC LOBE?

". . . it represents a primary division more basic than any lobe; in addition, it actually includes several lobes, hence the word lobe does not sufficiently characterize it. ***Therefore I call it the great limbic lobe****."*

(Broca, p. 394, Broca's emphasis). Trans. by Schiller, p. 258, 1992.

Broca's decision to embrace a rather robust limbic lobe construct, extending out to the limbic fissure (including the entire gyrus fornicatus, and thus

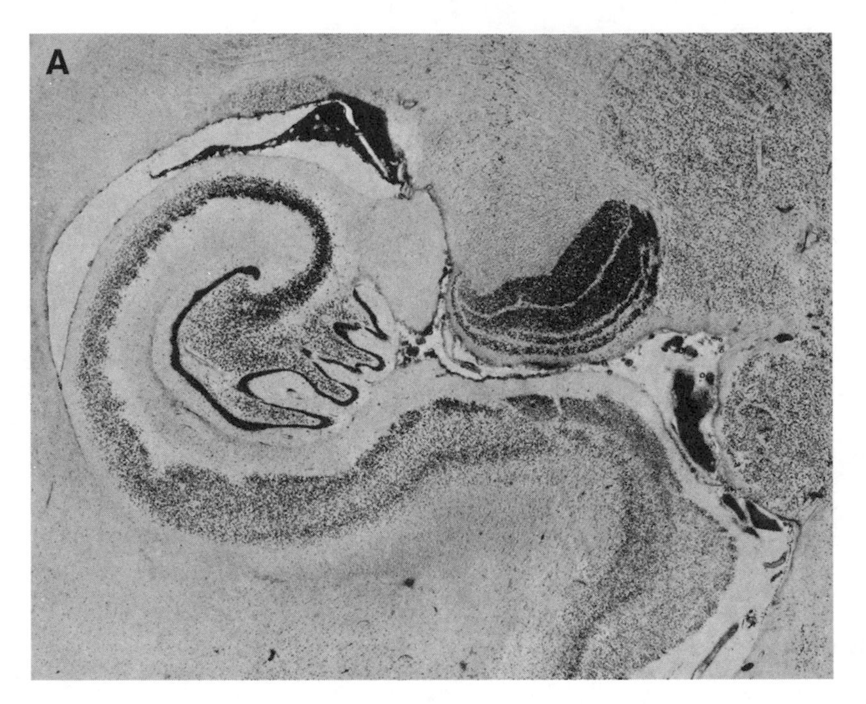

Figure 1.7A. Cross-section of the temporal lobe at the level of the geniculate nucleus in the adult. Note the salience of the cortical limbus, and compare with figure 6, the gray white relationships remain essentially unchanged between embryo and adult. The diagram reveals rather strikingly the spiral pattern mapped out by the parahippocampal gyrus, the hippocampus, and dentate gyrus. The cortical limbus stands in close proximity to the central axis of this spiral. Figure reprinted with permission from (16). The remaining four illustrations in the figure 1.7B–E demonstrate the behavior of the continuous line of cortex at the surface of the brain from the rhinal sulcus to the corticoepithelial border. The principle deformations (folding) of this "line" of cortex are delineated. In 1.7D, Gloor, has (for illustrative purposes) "opened up" the hippocamplal sulcus, compare 1.7D (schematic) with 1.7E (also 1.7A).

extending well beyond the true cortical—gray/white—border, see figure 1.7) places him in a school of 19th century neuroanatomists who, not constrained by conventional lobar anatomy, seemed more interested in large scale hemispheric design principles. The conceptual alliance of this school (as exemplified by such constructs as Gerdy's *circonvolution annulaire*; Fovilles' *grande overt de l'hemisphere;* Meynert's, also Foville's concentric ring model, and Broca's great limbic lobe) with Geoffroy Saint-Hilaire's transcendental morphology has already been mentioned, also see Table 2.

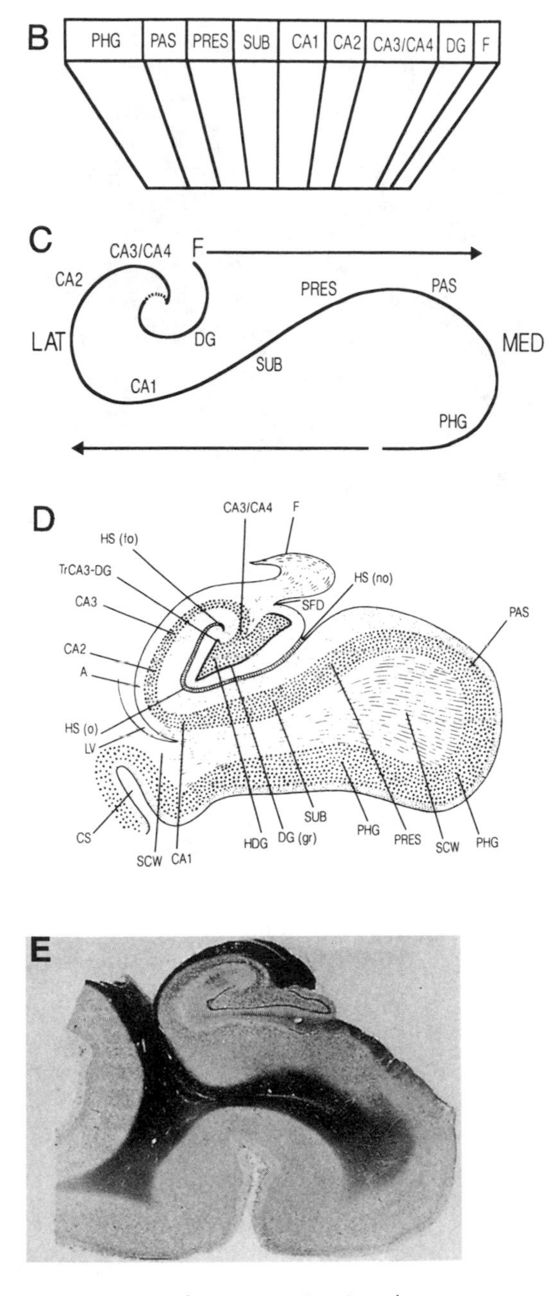

Figure 1.7. *Continued.*

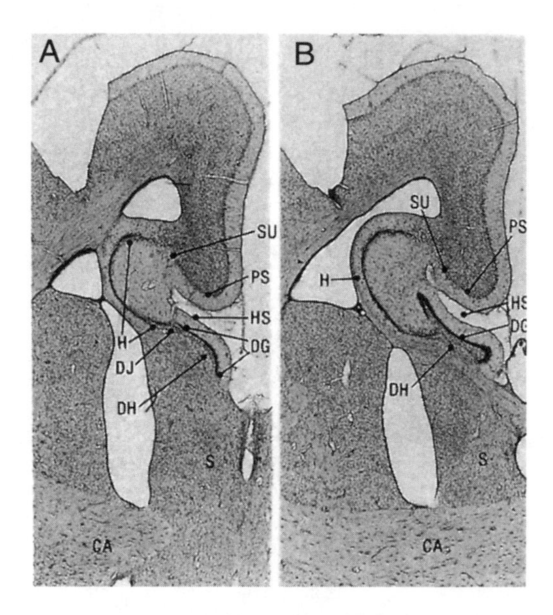

Figure 1.8. A) Coronal section through the hippocampal region of the opossum, (A) is more rostral to (B). The continuity of the medial pallium throughout the hippocampal formation is clearly displayed in (A). In (B) the continuity of the hippocampus (Ammon's horn) is lost. Reprinted with permission from Gloor (9).

AN EXPLORATION OF THE HIPPOCAMPAL REGION/CORTICAL EPITHELIAL BORDER (CORTICAL LIMBUS) IN THE HUMAN TEMPORAL LOBE

"The hippocampal region, defined as the area that extends from the collateral sulcus to the epithelial margin of the hemisphere marked by the choroid plexus is a continuous sheet of cortex folded in opposite directions in three consecutive turns. The folds become tighter as one approaches the epithelial margin of the hemisphere."

(9, p. 326)

The cortical epithelial border, the cortical limbus, is a canonical element in the tale of limbic brain anatomy. Broca appears to have well appreciated it referring to this border as the opening into the inner recesses (lateral ventricles) of the hemisphere, *As it surrounds the threshold of the hemisphere the mantle forms a border which resembles the circular edge of a purse. Hence I am calling this border the limbus of the*

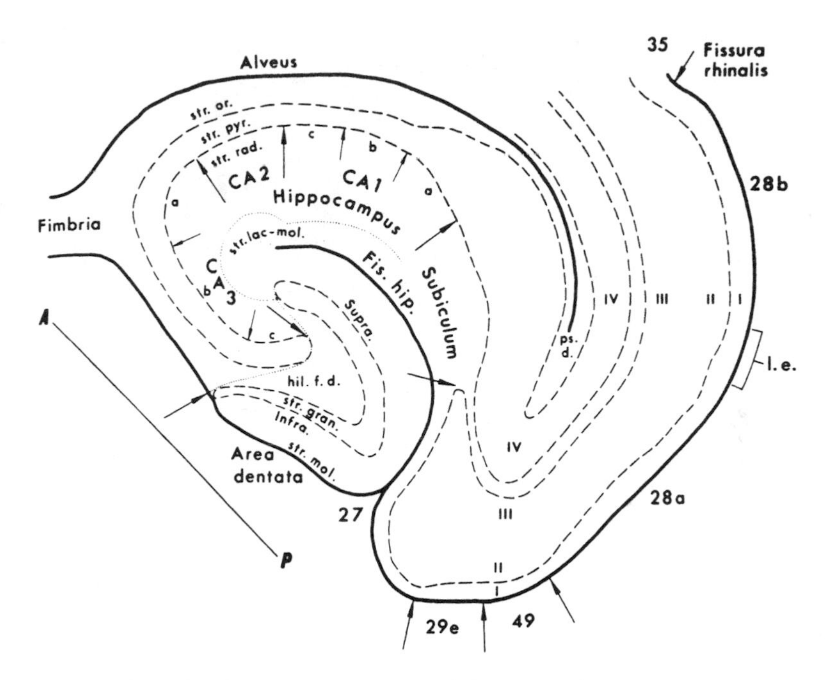

Figure 1.9. Schematic of hippocampal region of the mouse, horizontal section. Note that between the rhinal sulcus laterally and the presubiculum boundary medially the cortical plate separates into a two layered arrangement (the outer plate is cortical layers II-III; and the inner plate layer IV: the cortical plates are separated by a cell free zone, the lamina dissecans). This double layered expanse marks the outer (mesocortical) ring of Broca's lobe. At the presubiculum/subiculum boundary the outer cortical layer attenuates and then disappears: in earlier writings this is referred to as the limes duplex of Filiminov. Reprinted with permission from Angevine (17).

hemisphere, and the convolution that forms it the "limbic convolution" (Broca). And recall Nauta's admiration of Broca's exploration and discovery in this neural terrain. *Little wonder, all in all, that Broca* decided *he had found the edge of the cerebral cortex, at least in the temporal lobe, and little wonder that he initially chose to call his discovery the great lobe of the hem* (Nauta).

We must perforce draw our attention to this anatomical terrain and observe this limbic border first hand. The anatomy of this area is daunting, we therefore restrict the description of this area to fundamentals (exceptional treatments of this complex anatomical landscape have recently become available, e.g. Gloor's recent text (9), also Duvernoy's atlas (14)).

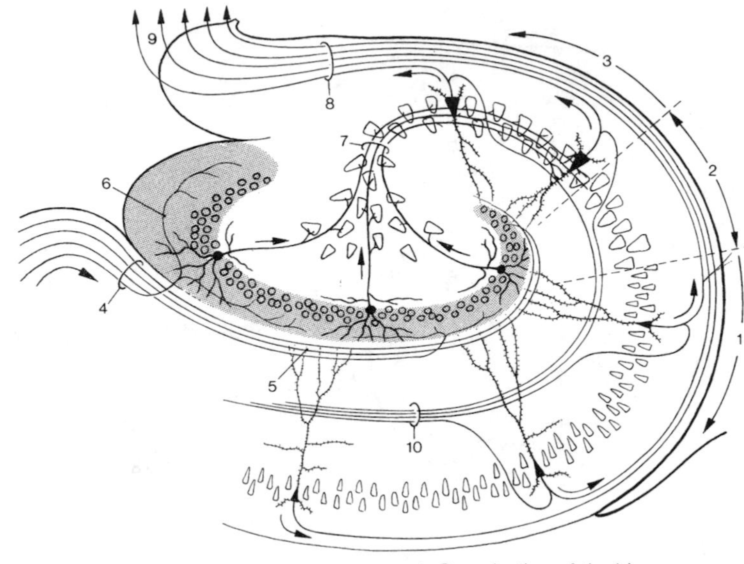

Figure 1.10. Schematic represents the basic connections of the hippocampus. Note how the afferent fibers synapse on the single layered granule cells. Also note how that in the dentate gyri, at the end of the cortical gray, the granule cells appear to perched (only partially overlapping) the neighboring (CA 3) hippocampal pyramidal cells. The granule cells are afferent elements of the hippocampus, their apical dendrites are oriented toward the pial surface. The granule cells in turn send their axons as mossy fibers, to the efferent pyramidal cells. This circuitry is illustrative of the basic two-neuron reflex, but positioned at the level of the cortex. Figure from (3, p. 221).

Three strategies are employed to help render basic concepts of hippocampus and corticoepthelial (limbic) border anatomy more accessible to the reader new to this area.

1) Attention will be directed at a cross sectional of the *body* of the hippocampus. There are two reasons fort this. a) A cross section of the temporal lobe through the body of the hippocampus displays both the full double-C cytoarchitectonic profile of the hippocampal region, and the corticoepithelial margin. b) Cross section of the head of the hippocampus introduce complexities beyond the scope of an introductory treatment.
2) The anatomic survey will be divided into two parts. The first part reaches from the rhinal\collateral sulcus to the fimbriodentate sulcus. The second part extends

from the fimbria white into the hemispheres midline cabling systems. This division is, in large part, a false dichotomy to be sure, but it is employed to help better introduce fundamental relationships encountered at the cortical limbus.

3) In the first part of the survey we will track a *continuous* line of cortex to serve as a guide or pathway from the rhinal sulcus to the dentate fimbria border. Reference to schematic illustrations will be made to render this *continuous* cortical pathway readily visible throughout this neuronal expanse.

FROM THE RHINAL/COLLATERAL SULCUS TO HIPPOCAMPAL SULCUS TO FIMBRIODENTATE SULCUS

Figure 1.7A, a cross section at the hippocampal body provides the aforementioned desirable features enabling an introductory survey of cortical limbic anatomy. Although the fimbriodentate *sulcus* (the end point for the first leg of this journey) is not marked in figure 1.7A, it is easily discerned. The fimbriodentate sulcus is the space between the last "tooth" of the dentate gyrus and the fimbrial white: this sulcus separates the fimbria, located above this sulcus, from the dentate gyrus, located below this sulcus. It should be appreciated that the dentate gyral tissue located just under the fimbriodentate gyrus is visible on the medial surface of the temporal lobe, it is referred to as the margo denticulatus. Although, figure 1.7A fails to include the collateral sulcus, it would be positioned at or near the cut off (in this figure) of the parahippocampal gyrus.

Figure 1.9 represents the same neural expanse (rhinal sulcus to dentate fimbria border) in the mouse brain. Figure 1.6 is included to demonstrate cortical limbic relationships as depicted in an embryonic schematic. Gloor's Schematics (figures 1.7B–1.7D) provide a rendering of the continuous cortical line (not visible in figure 1.7A). With these illustrations, we begin our journey.

Collateral Sulcus to the Entrance of the Hippocampal Sulcus: The Parahippocampal Sectors (The Mesocortical Sectors)

The parahippocampal gyrus "begins" on the medial surface of the collateral sulcus, and "ends" (as the presubiculum) on the lower lip of hippocampal sulcus. The crown of the parahippocampal gyrus, which is fully exteriorized, forms the medial edge of the temporal lobe: it forms the inferior temporal arm of Broca's great limbic convolution.

The parahippocampal gyrus is so named because it lies along side (para) to the hippocampus. (The term gyrus hippocampi was used by older anatomists to refer to the parahippocampal gyrus, (9, p. 52); use of either, gyrus hippocampi, or equivalently hippocampal gyrus to refer to the parahippocampal gyrus, leads to confusion (with the hippocampus, an entirely different structure); therefore the term hippocampal gyrus is to be avoided. The mesial wall of the collateral sulcus (where the parahippocampal gyrus begins) is constituted (in rostral sectors) by entorhinal cortex.

Entorhinal cortex refers to its position medial to the rhinal fissure (see figure 1.9 for representation of the rhinal fissure in the mouse brain). Entorhinal cortex is considered, on cytoarchitectonic grounds, to be *meso*cortex—cortical terminology and cytoarchitectonics are briefly introduced later in this chapter. Entorhinal cortex, on topological grounds, is considered to be *medial* mesocortex (as opposed to lateral mesocortex). The *medial* mesocortex is derived from the *dorsal* pallium, see figure 1.11 (also see Gloor (9) Plate II). The band of medial mesocortex, of which the entorhinal cortex is just one part, runs alongside structures derived from the *medial* pallium (the medial pallium lies, as expected, medial to the dorsal pallium). We will encounter the structures derived from the medial pallium later in this examination. The terminology of cortical topology (i.e., the medial, dorsal, and lateral pallium) is introduced later in this chapter. Put simply, the medial mesocortex is sandwiched between two layers. Medial to the medial mesocortex is a layer (not yet encountered) derived from the so called medial pallium. On the other (lateral) side of he medial mesocortex is neural tissue derived from the dorsal pallium—the medial mesocortex is also derived from the dorsal pallium.

Proceeding in the direction of the cortical limbus, the next cortical divisions encountered are the parasubicular respectively presubicular divisions of the parahippocampal gyrus (9, p. 326). Their cortical investment is medial mesocortex (we are still in the area of the cortical plate derived from the dorsal pallium). The presubiculum stands at the entrance to (and forms the lower lip of) the hippocampal sulcus. The upper lip of the entrance to the hippocampal fissure is formed by the dentate gyrus (figure 1.7A). The interested reader should compare the position of the lips of the entrance to hippocampal sulcus in figure 1.7A (human) with figure 1.9, (mouse brain). In the mouse brain, complete rotation of the hippocampal region is not exhibited: what forms the lower lip of the hippocampal sulcus in the fully rotated, temporalized human brain (figure 1.9), forms the upper lip in the non rotated rat brain. Notably, the *entrance* to the hippocampal fissure is fully patent in both the human and mouse brain: indeed it is patent in all mammalian brains.

A Survey Along the Entrance to the Hippocampal Sulcus: Transition From Presubiculum (Mesocortex) to Subiculum (Allocortex). Transition from Two Tiered Mesocortex to a Single Tiered Allocortex

> The division of the cortical ribbon into these two principal layers or broad filaments is evident throughout the entire posterior and posteromedial curvature of the cerebral dhemisphere [Angevine is referring to the mouse brain, see figure 1.9]. Pereception of these two filaments is an important first act in approaching the troublesome anatomy of this part of the cerebral cortex.
>
> 17, p. 64

The presubiculum extends for a short distance beyond the opened, non-obliterated sector of the hippocampal fissure (figures 1.7, 1.8, 1.9), where it is soon replaced by the subiculum. The transition from the presubiculum to the subiculum represents, on cytoarchitectonic grounds, the transition from mesocortex to allocortex. The transition from the presubiculum to the subiclum also represents, on topological grounds, the transition from dorsal pallial derived structures (and these structures are the already discussed parahippocampal divisions) to structures derived from the medial pallium: i.e. subiculum, hippocampus, and dentate gyrus.

Because it easy, even in an introductory treatment of the hippocampal region, to be quickly overwhelmed by terminology, in the paragraphs that follow attention will be given to more global aspects of the anatomy.

One of the most readily recognized features of the neural terrain we are scrutinizing is the transition from thickly layered cortex of the parahippocampal gyrus to the single tiered much thinner cortical investments of the subiculum, hippocampus, and dentate gyrus (figures 1.7, 1.9 and 1.10; also see Nauta's legend to his figure 104—reproduced in this chapter as figure 1.7, without original legend).

Angevine, see his epigraph above, instructs us first to focus our attention on a particular aspect of this transition. Specifically Angevine asks us to observe that between the rhinal sulcus laterally and the hippocampal sulcus medially, the cortical plate has a two tiered appearance. This two tiered architecture is *not* seen outside these boundaries. The two tiered appearance readily evident "throughout the entire posterior and posterior-medial curvature of the cerebral hemisphere" in the mouse brain schematic, see figure 1.9, is also evident on inspection of the human brain, see figure 1.7.

The *outer* tier of the presubicular cortex abruptly stops at the presubicular subicular border, see figure 1.9 The subiculum is constituted only by the *inner* "deeper tier". Examination of this transition provides a window into nature's plan for elaborating multi-tiered cortical architecture. Gloor draws our attention to this, "if one follows the inner layer of the subicular allocortex towards its transition to the adjacent mesocortex, it appears as if the thick pyramidal layer [the inner tier] continues as an unbroken and rather uniform band into the deep layers of the adjacent pre- and parasubiculum and subsequently into layer VI of entorhinal (isocortex). It thus looks as if, from the presubiculum on toward the entorhinal cortex, new layers were piled on top of the subicular pyramidal layer." (9, p. 345).

Russian workers early in the early 20th century were among the first to appreciate how research into the architecture of the hippocampal region (with its evident transition from a two tiered cortical plate to a single tiered plate) could shed light on basic cytoarchitectonic and tectogenetic principles of hemispheric construction. The following description from Paul Yakovlev (1894–1983) acknowledges their efforts.

> The limbic isocortex [here Yakovlev is referring to the two tiered mesocortex], however, is sharply demarcated from allocortex [i.e. the single tiered subiculum] by a number of characteristic cytoarchitectonic features, the most definitive of these is the superlamination of the allocortex by the superficial layer of the isocortical [read mesocortical] plate, which terminates in a sharp clinoid border—"the limes duplex" of Filimminov (1938). This clinoid border and zone of superlamination *define the lip of the hemisphere in a strict tectogenetic and architectonic sense.*"
>
> YAKOVLEV, emphasis added (18, p. 77)

One can go on at some length regarding early preeminent descriptions of this anatomic terrain, (recall the Kappers comment, cited in the preface, "Certain of the most excellent contributions in to the history of comparative neurology have dealt with the hippocampal region of mammals"). In closing this section we simply mention one additionally "most excellent" description by Maximilian Rose (1927). Rose referred to the two tiered architecture as the laminae principales interna et externa, it is interesting to contrast his description with Yakovlev's (the reader is referred to the chapter by Angevine for further description of Rose's work).

Before closing this overview of the presubiculum to subiclum transition, it is to be noted that all three structures, including the parsubiculum are sometimes grouped together and referred to as the subicular complex this derives from the fact that these areas appear to undergird or suppor the hippocampus proper—reference to figure 1.7 readily affirms this appearance (subicere, from the French to support).

A Survey Along The Obliterated Hippocampal Sucus: The Hippocampus (Hippocampus Proper) Ammons Horn

Following the cortical plate along the hippocampal fissure, the subiculum is replaced by the hippocampus. The transition is not abrupt. Gloor writes that the hippocampus begins as the underlying white matter narrows to form the thin plate of the alveus that now separates the cortical gray matter of the hippocampus from the underlying temporal horn of the lateral ventricle. Angevine writes that (in the mouse) the transition occurs at the point where a crowded, densely stained layer of ammonic pyramidal cells suddenly appears along the superficial border of the stratum pyramidale of the subiculum (17). In Angevine's schematic the border between the subiculum and hippocampus is indicated by the arrow.

In the area of the subiculum, even before we reach the hippocampus, a problem is encountered which threatens to prevent our examination from proceeding further. Our journey is predicated on following the *continuous* line of the cortex from the rhinal sulcus to the fimbria white (see the 3rd of our simplifying strategies we made above at the start of section. This continuous cortical line, easily visible along the interhemispheric fissure and also the *entrance* to the hippocampal sulcus is obliterated beyond the entrance to the hippocampal sulcus.

Since the medial aspect/surface of the cortical plate, the medial pallium had to be continuous at the very earliest stage of neuroembryogenesis, an event must have occurred during embryogenesis, which obliterated the continuity of the cortical surface. This event was the increasingly tight spiral folding experienced by the medial cortical plate early in development. The spiral infolding caused the most medial aspect of the cortical plate (what becomes the dentate gyrus) to fold back into the adjoining, neighboring hippocampal sectors of the cortical plate: the dentate area in effect is whorled back into the hippocampal region, and in the process the hippocampal sulcus obliterated.

This problem creates difficulties not only for our journey, but for any student first encountering this part of the hemisphere. In most other areas of the brain there are well defined sulci and adjacent gyri to serve as reference: Imagine what it would be like to identify the pre and post central gyri if the central sulcus were nearly completely obscured (filled in perhaps by "bridging" neural tissues). Identification would obviously be more difficult. Yet, this is essentially the situation encountered in studying the hippocampus and dentate gyrus. Nature has hidden (filled in) the sulcus between them.

(Notably, although this filled in hippocampal sulcus makes learning the anatomy much more difficult, the collapse of the hippocampal sulcus *pari passu* the increasingly tightened hippocampal spiral may make neural afflux—via the mechanism of the perforant pathway[2]—to the dentate more easier (i.e. faster) by decreasing the distance traveled by incoming perforant axons).

Figure 1.6, and 1.8 are included to represent embryologic and phyletic/comparative analyses designed to better understand the closure of the hippocampal sulcus. Figure 1.6 is from His, and seems to represent an attempt to show patency of the sulcus at the 4 months stage of development. Figure 1.8, depicts two coronal sections through the hippocampal region of the opposum: at the rostral section in this precallosal mammal, the continuity of hippocampal and dentate cortex, and a patent hippocampal sulcus is evident.

Visualizing (Unobliterating) The Hippocampal Sulcus: Sectors CA1–CA3 on the Lateral Side of the Hippocampal Sulcus

In search of a visible hippocampal sulcus to map out our path, figure 1.7A, by itself fails us. Only the vestige of this sulcus, barely visible, is evident in the figure (the vestigial sulcus is made of fine blood vessels, leptomeningeal remnants, small CSF pockets, see (14, figure 7, also figure 8). A distinct cortical line has vanished.

A new map displaying the course of a "patent" hippocampal sulcus is necessary if we are to make progress here. Gloor's recent text provides just the schematic we are looking for (a perfect complement to figure 1.7A), see figure 1.7D. Indeed, in Gloor's schematic the hippocampal sulcus is depicted as patent and mapped all the way to its "obliterated" fundus. With a "patent" hippocampal sulcus as reference we can resume our journey, *along a continuous line of cortex*, to the corticoepithelial border.

Microscopic examination reveals the hippocampus, on the basis of different aspects of its *pyramidal* neurons, to be a heterogenous structure: four principal hippocampal divisions (CA1-CA4) are described. Hippocampal sections CA1–CA3 are sequentially arrayed along the *lateral* wall of the

[2] The perforant pathways coursing along the entorhinal area provides the hippocampus and dentate gyri with their most substantial input. The perforant pathway is actually visible as white matter, with appropriate staining, on the *surface* of the parahippocampal gyrus.

hippocampal sulcus: the term lateral here refers to its *topographic* position as viewed in figure 1.7A.

A Brief Review of the Territory Covered

Recall that we had proceeded through the opening of the hippocampal sulcus, the pathway we followed was along the *lower* lip of the entrance to this sulcus, the lower lip constituted by the presubicular, respectively subicular subdivisions. This lower lip becomes the lateral aspect of the hippocampal sulcus, constituted by the subiculum and CA1-CA3. Passing over these sectors takes us to the fundus of the obliterated sulcus.

Fundal Area of Obliterated Hippocampal Sulcus: The Zone of Transition from CA3 to the Dentate Gyrus

In an area where description of a continuous cortical line might seem most improbable (in the depth of the obliterated hippocampal fundus) a continuous line of cortex is discerned. Gloor explains, "There [and here Gloor is referring to the fundus of the hippocampal sulcus] *as in the depth of any sulcus, the cortex turns around its fundus*: the cortex ventro*lateral* to the walls of the obliterated hippocampal sulcus, which consists of hippocampal allocortex, turns around its bottom to form the *medio*dorsal wall formed by the dentrate gyurs. This coincides with a striking change in the histologic structure of the cortex." And just a few sentences later, Gloor continues, ". . . A rather loosely organized segment of the CA3/CA4 region, however, *links up* with the *edge* of the inner blade of the dentate gyrus." (9, p. 345, also see Gloor, his figure 5-1B, and 5-1C).

Notably, even though this part of the cortical plate (medial pallium) undergoes such intense folding/whirlpooling during development (compare figure 1.6 with figure 1.7), the behavior of the cortex as it rounds the fundus of the hippocampal sulcus (at the hippocampal to dentate cortex transition) is not essentially different from the behavior of cortex as it rounds (for example) the fundal depth of the central sulcus.

The CA3 to Dentate Gyrus Transition (Along the Surface of the Dentate Gyrus); and, the CA4 to Dentate Gyrus Transition (Into the Hilus of the Dentate Gyrus)

Under the influence of rotation, the cortex investing the depth of the hippocampal sulcus, follows two pathways after it turns around its fundus. 1) A rather loosely organized segment of the CA3/CA4 region links up with the edge of the inner blade of the dentate gyrus: this is Gloor's description previous cited (9, p. 345). This link provides the continuous cortical lining charting our exploration. 2) The bulk of CA4 (the CA4 section is referred to as the end blade) rather than *directly* continuing into the granular cell layer of the dentate gyrus, penetrates its hilus. This behavior of the end blade is described in Gloor's text, as a discontinuity, "There is an apparent discontinuity of the end blade of the hippocampus, which rather than directly continuing into the granule cell layer of the dentate gyrus appears to penetrate its hilus" (9, p. 345).

It would appear that there is a fan out of the brain parenchyma in the region rounding the fundus of the hippocampal sulcus. A small cortical surface segment of CA3/CA4 cortex bends most severely maintaining in tenuous form the original continuity of the cortical plate (9, p. 345). The bulk of CA4 the end blade of the hippocampus (CA3/4) bends less severely (but is still rotated in the same general direction) and flows into the *hilar* aspect not the *cortical* aspect of the dentate gyrus.

Dentate Gyrus to Fimbriodentate Sulcus

Before we extend our exploration to the dentate, we must dispense with a minor terminologic issue. The term hippocampus sometimes is used to include the dentate gyrus (9, p. 32). If an author feels the need to make a distinction, Gloor suggests using the term hippocampus proper (for CA1–CA4), in contradistinction to the dentate gyrus.

Gloor (see above) directed out attention to the striking change in the histologic structure of the cortex coincident with its turning around its bottom (fundus) to form the mediodorsal wall—the dentate gyrus. This striking change relates mainly to the appearance of the granule cell layer of the dentate gyrus. The granule cell layer of the dentate gyrus is so densely packed, that it appears as a black line outlining the entire course of the gyrus, see figure 1.7. Toward the termination of the dentate gyrus, as depicted in figure 1.7, the gyrus exhibits an erosed or tooth like pattern (hence its name). The last tooth-like serration of the dentate gyrus sits just under the fimbriodentate sulcus.

HIPPOCAMPAL TERMINOLOGY

> "The hippocampus is an outstanding example of what Burdach has called "Wit der Namengebung" (inventiveness of name giving)"
>
> Alfred Meyer, Historical Aspects of Cerebral Anatomy (19)

The hippocampus is an area of the brain, which has, over the centuries, been described with a truly bewildering variety of names.[3] Anatomists have perceived the area to resemble a silk worm, bombycinus, and of course a sea horse, hippocampus. Samuel Soemmerring (1755–1830) known to all medical students for what has been, over the last two centuries, the system used in naming of the cranial nerves, suggested the seemingly straight forward term gerollte Wulst (rolled in gyrus) for the hippocampus. (Incidentally, Soemmering's system for numbering the cranial nerves had its basis in the holes in the floor of the skull through which nerves extend out from the cranial cavity). Space limitations prevent any further digression into this engaging and colorful chapter in the history of neuroanatomy (20). However, the following, issues relevant to hippocampal terminology are essential and therefore require a brief discussion.

— Brodmann did not map beyond the presubiculum (see figure 1.9, the presubicular cortex is Brodmann area 27), therefore his illustrations do not serve as an aid to the hippocampus. The map in common use is from the histological analyses carried out by Lorente de No (a student of Ramon y Cajal). In his areal parcellation of the hippocampus, Lorente de No described four sectors, CA1–CA4.

— CA, (Cornu ammonis) was the term employed by Noguez (1726) and Garengeot's (1728) to refer to the human hippocampus. Possibly for reasons of appearance (longitudinal sections do reveal a ram's horn like architecture to the hippocampus) and also hubris (it is not unappealing to compare a part of the human brain with that of the Egyptian god who sported a ram-like head) the term Corne d'Ammon remained in use for two centuries, and thus available for use by Lorente de No in his histological studies carried out in the 1930's.

— Hippocampal Formation: Citing Gloor, the frequently encountered term *hippocampal formation* lacks a precise definition. The term refers in its most restricted sense to the hippocampus proper, and the dentate gyrus. Beyond this, the extent of the hippocampal formation, varies. In Gloor's text, the hippocampal formation is defined principally as the hippocampus proper, the dentate gyrus, and the subiculum (9, p. 52). However, see text below, one can make (and Gloor does) an argument to extend the hippocampal formation to also include the corpus callosum, and dorsal portion of the septum pellucidum (9, p. 79).

---

[3] A principal underlying (etymological) theme for the terms applied to areas proximate to the limbic border (hemispheric hilum) reflect the arc wise growth pattern mapped out by these structures. Such terms include Ammon's horn, hippocampus, cingulate (from the Latin for girdle) gyrus, falciform lobe (L. *falx*, sickle). Fornix, and gyrus fornicatus (L. fornix, arch or vault). The sexual connotation for this later term seems to refer to late night activities under archways in ancient Rome.

The Term Rhinencephalon

Most of the structures in the limbic are so named on the basis of morphology and it is not difficult to appreciate why they are so named, for example the cingulate, and fornicate gyri, fornix, Ammon's horn, hippocampus. However one term, rhinencephalon, it is not easy to trace its foundations. Because this term has been equated with Broca's limbic lobe (very much so in the past) a brief discussion of this term follows:

The term was first proposed by Etienne Geoffroy Saint-Hiliare in an unrelated context, to refer to a type of uniocular monster. Richard Owen (1804–1902) employed the term, which means nose brain, in a neuroanatomic context to refer only to the olfactory bulb and peduncle. Sir William Turner (1832–1916) extended its meaning to include the entire piriform lobe. Turner's publication of 1890 helped to popularize this term.

Elliot Smith's article (21) sheds light on the need for Turner (also His and others) to employ this term. Turner it seems wanted to describe basal hemispheric structures in a fashion complimentary to Reichert's concept of the hemisphere. Reichert put forth the term pallium (1859) to help in the description of the embryonic human cerebral hemisphere. The thin upper part of the embryonic wall he termed pallium (mantle) to be distinguished from the thicker basal mass (he called (*Stammlappen*).

Elliot Smith pointed out problems with the term pallium which in turn resulted in problems with the term rhinencephalon. The reader is referred to Smith's article for an introduction to this area.

It is more germane to the body of this chapter to note that after Turner publication (1890) the use of the term rhinencephalon gained considerably in popularity in England, and the continent. An even broader application of the term occurred in Europe (particularly in France) where the use of the rhinencephalon became conflated with Broca's concept of the great limbic lobe.

This rather expansive use of the term brought problems in its train. Particularly, as pointed out by MacLean, was that in the first half of this century, it acted to stifle interest in this area of the brain. MacLean continues that with the growing appreciation of man's microsmatic, plus improvements in hygiene the belief grew that the sense of smell and relatedly the nose brain was of diminished importance in human beings (5, p. 263). The term had been used as recently as the 28th edition of Gray's anatomy.

CORTICAL TERMINOLOGY

"The microscopic study of the cortex was in its infacy at the time of Broca. Subsequent studies revealed that most of the evolutionary old cortex is contained in the great limbic lobe."

PAUL MACLEAN, The Triune Brain (5, p. 259).

In the preceding discussion of the neural terrain from the rhinal sulcus to fimbriodentate sulcus, the terms used to refer to the cerebral cortex (iso-, mes-, and allocortex) were employed without elaboration. In the paragraphs that follow cortical terminology will be briefly overviewed. Familiarity with the basic elements of cortical terminology (and relatedly, cytoarchitectonics) is essential to a proper understanding of the limbic brain concept.

There are two major types of cerebral cortical tissues: the isocortex, and the allocortex (these terms were introduced in 1919). Isocortex forms the bulk of the surface of the human cerebral mantle. Isocortex appears in Nissl preparations to have six horizontal layers. Allocortex refers to cortical tissue other than (G. allos, other) isocortex. Allocortex differs from (is other than) isocortex by not having six layers. Allocortex is a thinner three layered cortex.

There are two types of allocortex. One of these types, already encountered in our discussion, is referred to as the hippocampal allocortex. The cortical investment of the hippocampal sectors CA1–CA4 is, of course, hippocampal allocortex. The dentate gyrus is also considered allocortex. Indeed, in the dentate gyrus, the three layer architecture (three layers are the defining feature of allocortex) are plainly visible (14, p. 8, also 14, figures. 7, 9). The hippocampal allocortex, as already discussed, is derived from the medial pallium.

The other type of allocortex is referred to as the prepiriform and periamygdaloid allocortex. The hippocampal allocortex and the prepiriform, periamygdaloid allocortex are brigaded together as allocortex, only because they share a three layered architecture: however, they, a) differ in their disposition/location on the hemisphere; b) differ in their function, and c) differ in origin.

The prepiriform, periamygdaloid allocortex is derived from the ventral portion of the *lateral* pallium. The *prepiriform* cortex is located along the lateral border of the lateral olfactory tract and reaches to the beginning of the pear shaped (pirifrom, formerly spelled pyriform: pirium = pear) bulge on the underside of the hemisphere (the bulge is formed by the head/uncal aspect of the hippocampus in the human brain: however, a piriform bulge is evident in less temporalized forms, see Gloor (9) his figure 2–15, also discussion by MacLean (5, p. 253–260)). The *periamygdaloid* allocortex (positioned just caudal to the prepiriform allocortex) surrounds the amygdala: this occurs in the head end of the hippocampus. Although of central importance to a more advanced understanding of limbic and hemispheric anatomy, this chapter will not include any further discussion of the prepiriform and periamygdaloid allocortex (as underscored above, our principal focus has been restricted to the body of the hippocampus, an area located caudal to the periamygdaloid allocortex, see figure 1.7).

Terminology of the Iso, Meso-, and Allocortex: Issues Related to Evolutionary Age

The term neocortex is often used interchangeably with isocortex. The term archicortex is often used interchangeably with hippocampal allocortex, and the term paleocortex is often used in place of prepiriform-periamygdaloid allocortex. The terms archi, paleo, and neo, connote quite strongly, differences in evolutionary age/origin (i.e. paleo means ancient) of the cortical investments.

However, in the introductory chapter to his text, Gloor points out 1) since *all* forms of mammalian cortex, including isocortex, have distinct premammalian precursors, and 2) because it is likely that cytoarchitectonic differences between iso- and allocortex arose very early in mammalian evolution, the descriptors (archi, paleo, neo) become unsatisfying, indeed misleading. What Gloor is saying, put more simply, is that if the neocortex (its precursor at least) has been around for nearly 200 million years, it can hardly be considered a new tissue, nor should it be considered a tissue appearing de novo, after the complete development of earlier (archi- and paleocortical) tissues.

PALLIAL TERMINOLOGY: MEDIAL, DORSAL, AND LATERAL PALLIAL DISTRICTS

> "The anatomical organization of the mammalian hemisphere, and in particular the relationship of its limbic structures with "non limibic" isocortex and some subcortical areas, can be more easily understood if one attempts to reconstruct the likely evolutionary pathway that led from an early vertebrate brain in which the prototypical topological design of the vertebrate telencephalon had remained relatively undistorted to the

mammalian and particularly the human brain where at first glance it is difficult to recognize its persistence."

PIERRE GLOOR (9, pp. 34–35).

In the preceding discussion of the neural terrain from the rhinal sulcus to fimbriodentate sulcus, the terms medial, dorsal, and later pallium[4] were used without elaboration. In this section, this terminology will be briefly overviewed. Familiarity with this description is also essential to a proper introduction to the limbic brain concept.

---

[4] The term pallium requires elaboration. This term originated in the mid 19th century with Reichert (1859). Textbook authors (e.g. MacLean, Gloor) seem to cite Elliot Smith's interpretation for what Reichert meant by his term. Elliot Smith (21) wrote, "Reichert [more than forty years ago] came to the conclusion that it would facilitate the accurate description of the cerebral hemisphere if the thin upper part of the walls of the embryonic cerebral vesicle, which he termed "pallium" (mantle) were distinguished from the thicker basal mass which he called the "Stammlappen." The later expression was so employed to include locus perforatus anticus, corpus striatum . . .".

More specifically Reichert's term pallium (according to Elliot Smith) "was employed to designate that cortical area (with its associated medullary layer) which is free from (i.e., is not adherent to the surface of) the corpus striatum. As such the name is of little morphological value." One of the immediate problems encountered with Reichert's term was the presence of cortex with a medullated layer found along the *basal* surface of the telencephalon.

Following the usage of the term pallium is clearly an exercise in the history of neuroanatomy too exhaustive to pursue here (the interested reader is referred to two rather extensive foot notes (9, p. 28, fn; also Kuhlenbeck, p. 472, 476) for an exploration of the many issues and problems raised with the term pallium). However, two historical aspects of the term 'pallium' are briefly noted below because they relate to the broader issue of the limbic brain concept.

a) In the same article referred to above, in which Elliot Smith (1901) both explained and criticized Reichert's term, he (Elliot Smith) introduced the term neopallium (neocortex) for the great progressive cortical field seen in mammals. Smith comments, on his decision "But instead of selecting a new phrase such as '*pars crescens*,' I prefer to use the term neopallium, because the basis of this hybrid conveys some idea as to the meaning of the expression."

Elliot Smith's proposal of the then new term neopallium motivated the German anatomist and neurologist, Edinger, to propose the term archipallium/archicortex (for what Smith had called old pallium). At about the same time Ariens Kappers suggested the term paleopallium (paleocortex) to refer to basal olfactory areas, the piriform cortex (MacLean, p. 254).

Several more new terms were born in the wake of Elliot Smith's description of the neopallium. Raymond Dart (Australian anatomist, a student of Elliot Smith) described the identification of two types of neopallium in the reptilian brain, one of which he called "parapyriform" and the other "parahippocampal". These layers were identified as intermediate respectively between the older paleopallium and neopallium (the parapyriform layer) and the older archipallium and neopallium (parahippocampal).

Figure 1.11. Standard textbook illustrations introducing the pallial and subpallial districts are either cross sections, or a combined cross and longitudinal sections, usually of the amphibian brain (see foot note 5 in text). This figure is Northcutt's representation of the left telencephalic hemisphere of supposed ancestral amphibian. The telencephalic hemisphere is divided into six major longitudinal phlyogenetic columns: three protoypic pallial and subpallial columns. Pallial districts are as follows: PI is a medial column, recognized as hippocampus. PII is subdivided into a medial and lateral component. The medial component of PII contributes to the medial mesocortical aspect of the limbic cortical ring. The derivative of the lateral aspect of PIII is recognized is piriform cortex. There are three subpallial districts giving rise to amygdaloid, striatal gray. As noted in the text, Gloor's modification of Northcutt (Gloor Plate II), also serves as a valuable illustration of basic pallial description. Figure reproduced with permission.

Figure 1.11 is R. Glenn Northcutt's (1969) oft reproduced schematic illustration of the left telencephalic hemisphere of a hypothetical ancestral vertebrate brain (amphibian stage[5]). Reference to this diagram will serve to introduce cerebral wall terminology. A modified rendering of Northcutt's diagram in Gloor's recent text (Plate II, Gloor chapter 2 is exceptionally helpful). In the discussion that follows we will refer to both figures 1.11, and direct the readers attention to Gloor's Plates (II also III–VII).

These terms (all following in the wake of Elliot Smith's neopallium) and their associated description (the terms referring to older and newer cortical layers with intermediate cortical layers interposed) led to (to paraphrase MacLean) the germ of the concept of "successive waves of circumferential differentiation" as developed Abbie, and later elaborated upon by Sanides in terms of the concentric growth ring hypothesis. Growth ring theory (a theory which sheds much light on cortical topology) is particularly helpful in understanding basic limbic lobe architecture. Growth ring theory in the context of the limbic lobe will be discussed in the body of this chapter.

b) Reichert's sense of the term pallium also influenced Turner's development of the term rhinencephalon, albeit in a complex way: Elliot Smith in the article referred to above, suggests that Turner employed the term rhinencephalon as a complement to Reichert's term. With Turner's publication (1890) the meaning of the term rhinencephalon grew increasingly closer to Broca's anatomic description; also with Turner's 'memoir of 1890' the term rhinencephalon gained increasing popularity throughout England and Europe: this popularity would last to the middle decades of the 20th century and acted to deter academic interest in this area of the brain (see Side Bar: Rhinencephalon).

[5] The importance placed on the amphibian brain as a model for the vertebrate brain prototype merits a brief historical digression.

In the early decades of the 20th century neuroanatomists conducting comparative, and histologic studies had produced maps of the hemisphere, which distinguished several distinct cortical, also subcortical divisions. (Representative of these advances were a celebrated paper by the Australian neuroanatomist Grafton Elliot Smith (1871–1937) proposing the natural subdivisions of the hemisphere (21) also the cytoarchitectonic maps of Korbinian Brodmann (1868–1918)).

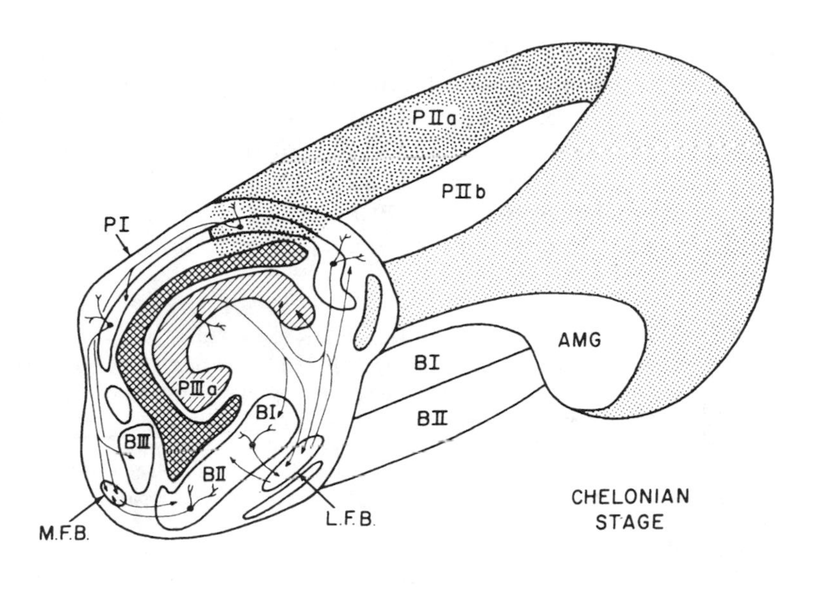
PIIa
PIIb
PI
AMG
BI
BII
PIIIa
BI
BIII
BII
M.F.B.
L.F.B.
CHELONIAN
STAGE

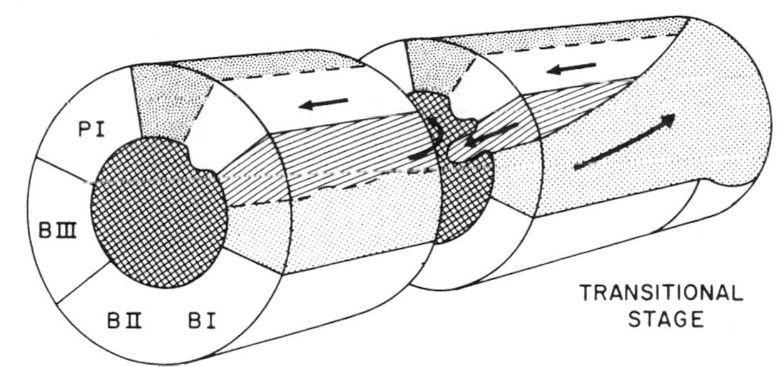
PI
BIII
BII BI
TRANSITIONAL
STAGE

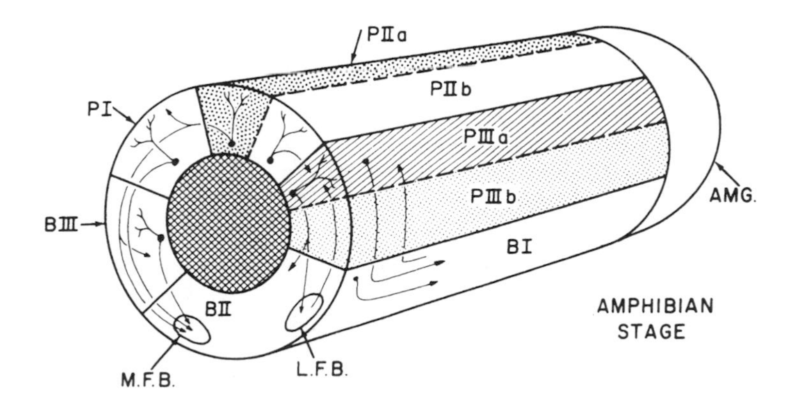
PIIa
PIIb
PI
PIIIa
PIIIb
AMG.
BIII
BI
BII
M.F.B.
L.F.B.
AMPHIBIAN
STAGE

Northcutt's schema posits six longitudinal columns. The longitudinal columns represent homologous structures in the vertebrate series. Phylogenetic/evolutionary fates of the columns are examined in Northcutt's article. Notably, the topological relationships of the columns (pallial districts) are preserved in all vertebrate brains.[6] Column PI (Northcutt) is the medial pallium. Columns PIIa and PIIb represent the dorsal pallium. Column PIIIa and PIIIb represent the lateral pallium. Columns BI–BIII are

---

Three principle hemispheric districts had been portrayed; these included: —) a medially positioned hippocampal formation; —) just adjacent to the hippocampal formation a somewhat more stratified and wider swath of cortex, the parahippocampal cortex; and —) the positioned more dorsolaterally a more well stratified district, the isocortex. Subcortical districts described were the amygdaloid complex, and corpus striatum.

It became a goal of comparative neuroanatomists throughout the early and middle decades of the 20th century to determine how these newly described pallial and subpallial subdivisions could be understood in a comparative analytic perspective. To paraphrase Kuhlenbeck, the emerging maps of the hemisphere posed additional basic problems related to a determination of morphologic homologies for the newly described telencephalic grisea (hemispheric districts) throughout the entire vertebrate series (23, p. 473). In this regard comparative neuroanatomy engaged the search for the manifestations of the homologies: the goal, as expressed by Kuhlenbeck, to permit a tentative tracing of the evolution of the telencephalon from fish to mammals (23, p. 471).

Early on in this effort it was recognized that one animal phylum, the amphibia, offered particular advantages for fundamental studies.

> The brains of some amphibians, . . . if perhaps not in all respects not the simplest vertebrate brain does indeed display the most clear-cut manifestation of the fundamental organizational pattern characteris of all vertebrates . . . , in particular it provides a most suitable frame of reference for an understanding of both comparative anatomy and presumptive phylogenetic evolution of the vertebrate forebrain."
>
> Paraphrase of Edinger (23)

The fundamental pallial divisions are *not* evident in the fish brain. These subdivisions began to emerge clearly only in tetrapods. It is in amphibians that the telencephalic divisions are exhibited in a particularly simple and easily distinguishable (on structural grounds) fashion (Edinger's point made above).

The opportunity afforded researchers by the amphibian brain would produce by mid-century such classics as Herrick's, *The Brain of the Tiger Salamander* (1948). Herrick's book provided a principal reference for illustrations of the basic hemispheric divisions (see this text, figure 2.2); also see Gloor, his figure 2.4 (9. p. 31); and of course Northcutt's illustration, figure 1.11.

[6] A most important aspect of understanding vertebrate brain anatomy across the series of all vertebrates (of tracing of the evolution of the telencephalon from fish to mammals, as Kuhlenbeck put it) is an appreciation of the preservation of topology. Topology is that branch of geometry that studies those properties of figure that remain invariant under all conditions of deformation (9). The position (topography) of the pallial and subpallial sections change dramatically from fish to callosal mammal; however, the topology is preserved.

basal neighborhoods, equivalently the subpallial districts. Column BIII is the septal district. Columns BII and BI are the subcortical gray, amygdaloid striatal districts. The following discussion covers only the most elementary concepts.

The Medial Pallium

Three principal derivatives of the medial pallium are described: the subicular allocortex, the hippocampal allocortex (hippocampus proper) and the dentate gyrus. In amphibians and reptiles the hippocampal allocortex still preserves a linear flattened appearance. In *all* mammalian brains, the hippocampal allocortex, and dentate allocortex (hippocampal formation) display the fully developed characteristic double-C pattern (dentate gyrus curved around the hippocampal gyrus). Coronal sections through the acallosal mammal brain of the opposum reveal in rostral coronal sections (figure 1.8A), a hippocampus laid out flat on the medial surface of the hemisphere in a pattern reminiscent of non mammalian amniotes (reptiles); and in more caudal coronal sections the mammalian double-C configuration (figure 1.8B).

The Dorsal Pallium

Running alongside (just lateral to) the medial pallium is the medial aspect of the dorsal pallium. The medial aspect of the dorsal pallium gives rise (in mammals) to the medial mesocortex.

The canonical midsagittal representaton of Broca's great limbic lobe (erstwhile referred to as Roland's processo cristato, Schwallbe's faciform lobe, or more commonly referred to as the cingulate, parahippocampal gyri) are constituted primarily by medial mesocortex (a cortical investment derived from the medial aspect of the dorsal pallium).

The Lateral Pallium

The ventral aspect of the lateral pallium gives rise to the olfactory cortex[7] (i.e., the prerepiriform and periamygdaloid cortex). The prepiriform cortex follows the lateral border of the lateral olfactory tract from the

[7] In amniotes, primary olfactory projections are now understood to be restricted to the *ventral* aspect of the lateral pallium. This in an interesting point in terms of the history/development of the limbic brain concept). Broca acknowledged a principal olfactory function for his great limbic lobe (although he did suggest an alternative function other than olfaction, and he was aware of microsmatic forms with large cingulated gyri). Also Cajal was unable to map direct olfactory connections to the hippocampus. None the less it became the predominant

posterior orbitofrontal region to its termination with the periamygdaloid allocortex. The tennis racket handle of Broca's great limbic lobe (the olfactory stalk) is a derivative of the ventral aspect of the lateral pallium As noted above we have elected not to examine the head end of the hippocampus in the chapter, this also precludes a description of the disposition of the periamygdaloid allocortex.

A more *dorsal* sector of the lateral pallium is described which gives rise to the *lateral* mesocortex (9, p. 18, & p. 33). The lateral mesocortex borders the prepiriform and periamygdaloid cortex.

Medial, Dorsal, and Lateral Pallial Districts: Brief Introduction to Hodology

In the preceding section, the pallial districts were defined, their disposition along the telencephalon examined, and hemispheric topology was touched upon. In this section we very briefly introduce hodologic principles as they relate to the pallial and subpallial divisions. Of course separating topological and hodological reasoning is a contrivance (a false dichotomy to be sure) however, in the service of a brief introduction to basic limbic anatomy we have taken this didactic license. Hodology refers to the study of neural connections. (*Hodos*, Greek for path or road, i.e. an electrode is a pathway for electrons; a neuronal axon is a pathway for ions: notably the term ion, also from Greek, means traveling).[8]

Medial Pallium

The medial pallium receives its principal input via its lateral edge from the neighboring district of the dorsal pallium (this is the anlage of the entorhinal perforant pathway). Outflow fibers from the medial pallium are positioned in the medial wall of the hemisphere. The output from the medial pallium either passes to the neighboring dorsal pallium (thus reciprocating

---

view point up until the middle of this century that Broca's limbic lobe was principally an olfactory brain (a rhinencephalon in the functional sense). Broca's reservations against his limbic lobe as purely an olfactory mediator, were proven correct. We can now state in the language of pallial hodology and topology that primary olfactory radiations are restricted (particularly in mammals) to the ventral pallium, and do not directly innervate medial pallial structures.

[8] It should be appreciated that Broca's concept of the limbic lobe was a large scale morphologic description which did not (and could not, given the technological limitations of the time) address the issue of neural connections nodology.

this input channel) or the output descends to lower stations of the neuraxis (septum, preoptic area, hypothalamus, midbrain).

The projection pathways to these lower areas of the neuraxis form classical limbic circuits: such as the fimbria-fornix system of mammals—this is the major pathway connecting the hippocampal formation with the septum, preoptic region, and hypothalamus; Nauta's limbic midbrain circuit (discussed in chapter 4); also the ventral amygdalofugal pathway[9] (this pathway has not been discussed in this chapter).

The anatomic behavior/disposition of these three pathways exemplifies a basic principle of limbic brain connections. Gloor explains. "What distinguishes the connections of the limbic structures from those of the isocortex is that they largely shun the internal capsule and descend to their subcortical targets . . . through the medial wall of the hemisphere."

Dorsal Pallium

An understanding of the basic connections of the dorsal pallium first requires an appreciation of the projection systems of the dorsal pallium in the prototype, and second an appreciation of how these connections vary between the ancestral model, and more elaborated forms.

In the ancestral state, the dorsal pallium employs two bi-directional fiber systems, one in the medial aspect of the hemisphere, the other in the lateral aspect of the hemispheric wall. The midline fiber system in its passage out of the dorsal sector joins the fibers exiting the medial pallium establishing connections with basal forebrain (septum, preoptic area) and more caudal brain stem stations. The lateral projection system engages the thalamus, and striatum, and lower brainstem sectors. In primordial forms, there is indiscriminate use (9, p.32) of both fiber systems by dorsal and medial pallial sectors.

In the brain of the reptile, also the mammal, the derivatives of the more laterally placed areas of the dorsal and dorsolateral pallium are elaborated as the reptilian dorsal cortex/dorsal ventricular ridge and the mammalian isocortex respectively. These laterally positioned derivatives in both the reptilian and mammalian brain restrict their association with the medial hemispheric wall/medial fiber system, employing predominantly the laterally positioned fiber pathways (i.e., the lateral forebrain bundle/internal capsule). The more medially placed sectors of the dorsal pallium, the

[9] Strictly speaking, the amygdaloid nuclei are not derived from the medial pallium, but subpallial sectors.

portion of the dorsal pallium flanking the medial pallium tend to retain their subcortical projections through the medial hemispheric wall, and tend to lessen their connections through the lateral forebrain bundle/internal capsule.

In the preceding paragraph, we see, in part, how the parcellation between the limbic and non limbic pathways evolves. The dorsolateral pallium gives rise to isocortex (mammals) and dorsal ventricular ridge (reptilian homologue of the isocortex). The isocortex preferentially engages the internal capsule, and disengages from the midline hemispheric white matter mechanisms—this is prototypical for non limbic circuitry. Meanwhile, the medial aspect of the dorsal pallium (similar to the medial pallium) projects to subcortical structures via midline fiber tracts—this is prototypical for limbic circuitry.

One last comment is merited. Two of the three limbic pathways discussed above, Nauta's limbic midbrain, and the ventral amygdalofugal pathway (mentioned above) project rather directly through the midline area to subcortical centers. Notably, the fimbria in the *reptilian* brain also passes its fibers directly to septal areas. However, this is not true for the fimbria fornix system of higher mammals, which exhibits a remarkable ring-like distortion. Why is this?

The distortion occurs because of the expansion of the hemisphere (growth of the thalamus, internal capsule, and the elaboration of the isocortex) in higher mammals. A discussion of the dynamics of mammalian hemispheric expansion are beyond the scope of this chapter (this area is handled rather expertly in Gloor's text). However, the long semicircular route (ring-like distortion) of the fimbria-fornix system (a principle limbic cable) also the stria terminalis (the stria terminalis connects the amygdaloid nucleus of the limbic system with the septum, preoptic region and hypothalamus, it is a dorsal outflow pathway of the amygdala) merit further comment. Therefore we must briefly consider the topic of mammalian hemispheric expansion.

The *primordial* rostral and caudal poles of the vertebrate brain, which are found at the rostral and caudal meeting points of the medial and lateral pallium maintain their basal midline positions, and their proximity to each other, throughout the vertebrate series. The midline rostral and caudal poles are represented diagrammatically as the septum, preoptic area (rostral pole) and the amydgala[10] (caudal pole).

[10] The amygdala itself is a sub pallial structure. The 'actual' posterior primordial pole is where periamygdaloid (lateral pallium) and temporal tip of the hippocampus (medial pallium) abut (9, p. 82). However, the underlying amygdaloid nucleus is said to represent the posterior pole (9, figure 2.21).

Importantly, the fornix system is "anchored" or "tethered" to these poles. During the isocortical growth (hemispheric expansion) seen in mammals, the fornix is bowed out from these anchor points. The isocortical expansion is likened to the cap of a mushroom. The collar around the stalk of the mushroom represents the midline limbic structures. The limbic structures are more closely tied to the fixed anchor points of the hemisphere than the isocortex. The isocortex, and its associated structures (internal capsule, also the dorsal thalamus) freely expand in all but the ventral (basal hemispheric) direction. Limbic structures such as the foures end up stretched around the dorsal mushroomed contour of the expanded internal capsule and assume a semicircular shape.

MacLean provides an evocative description of this process as it relates to the stria terminalis (but it would also relate to the fimbria fornix) of the higher mammal. "Because of its arc-shaped path, the stria might be viewed somewhat like a bridge with its supports embedded on the opposite shores of a river, forming a span from the amygdaloid nuclei on one side to the septum on the other side." (5, p. 288).

CORTICAL/HIPPOCAMPAL TERMINOLOGY: IMPLICATIONS FOR THE GROWTH RING HYPOTHESES

Interposed between the allocortical districts and the isocortex is a belt of transitional cortex. The transition from the three layered allocortex to isocortex is gradual.[11] The transitional cortex has been given different names, such as juxta or periallocortex, paralimbic cortex, and mesocortex.

This architecture compels a brief discussion of the growth ring hypothesis. As noted in the second foot note in this chapter, in the wake of Elliot Smith's influential publication introducing the term (and concept) neopallium, their quickly followed the term archipallium (Edinger), and paleopallium (Kappers). Archipallium referred to Elliot Smith's old pallium (our hippocampal allocortex). Paleopallium referred to the basal olfactory areas, the piriform cortex (our prepiriform-periamygdaloid cortex). (5, p. 254).

[11] In terms of our previous discussion of hemispheric pallial architecture, Gloor puts it thus, "All along the rostro-caudal extent of the hemisphere the dorsal pallium is intercalated between the medial and lateral pallium except at its rostral and at its caudal pole where the medial and lateral pallium come into opposition". (Gloor paraphrase, p. 36).

Particularly important in the development of growth ring theory was the identification by Raymond Dart[12] (former student of Elliot Smith) of two kinds of neopallium in the reptilian brain. One of these he referred to as parapyriform and the other parahippocampal. The parapyriform layer was sandwiched between the older paleopallium and newer (neo)pallium; and a parahippocampal layer was sandwiched between the older archipallium and neopallium.

Paul MacLean writes that with this description by Dart, was born the germ of the concept of "successive waves of circumferential differentiation". MacLean (5, p. 255) explains, "... Abbie, in line with the original observations of Dart, developed the concept that the cortex evolved by 'successive waves of circumferential differentiation'. Basing his argument on histological studies of the marsupial and monotreme brain, he [Abbie] proposed that the borders of the piriform and hippocampal area provided the starting lines of successively evolving cortical areas". Friederich Sanides further developed Abbie's theory of circumferential growth. Sanides called this process of cortical elaboration, the growth ring hypothesis[13]. Sanides, working in primates, identified many different transitional cortical districts (e.g., periallo- and proisocortex), which were interposed between the neo- and allocortex.

In terms of the growth ring hypothesis, Broca's circumannular convolution is composed of the two inner concentric rings of cortex. The inner most ring constituted by the allocortex (hippocampal, and prepirirform-periamygdaloid) and the outer ring by the transitional mesocortex. It bears reemphasis that Broca's limbic lobe construct was a gross anatomic description arrived at without any reference to cytoarchitecture.

THE HIPPOCAMPAL REGION: A SURVEY OF ITS COURSE THROUGHOUT THE HEMISPHERE

Earlier in this chapter it was stated that our examination of the hippocampal region would be directed at a cross sectional of the body of the hippocampus, and the reasons for this constraint were given. However, at this point we will go beyond the body of the hippocampus and survey the

---

[12] Raymond Dart discovered the first specimen of Australopithecus (*Australopithecus Africanus*) in 1925.

[13] The growth ring hypothesis has been subject to criticism, and it appears the theory is too problematic as a broad based theory of cortical evolution (24). Although of restricted applicability growth ring hypothesis remains helpful both as a partial description of mammalian cortical topology, and a window on limbic brain architecture. For an introductory discussion of the growth ring hypothesis in terms of limbic brain anatomy see, (5,24,25).

behavior of the hippocampus, cortical epithelial border in other hemispheric districts. There are two compelling reasons to extend our survey. a) The fundamental relationships expressed in figure 1.7 are maintained throughout the entire hemisphere—it couldn't be otherwise, because topology, the relationship of the pallial, and subpallial districts, is preserved. What we learnt in our examination of the body of the hippocampus applies throughout the other hippocampal sectors. b) A simplified geometry of the hemisphere suggests a spiral-like growth of the hemisphere around an axis drawn through the lateral fissure, see figure 1.12. This simple geometric perspective allows one to simply rotate figure 1.9 around a transverse axis drains through the hemisphere—this transverse axis often drawn through the amygdaloid region, or the limen insulae or passing through the lateral fissure, lenticular nucleus, mid IIIrd ventricle respectively, see figure 1.12.

Before we proceed any further, it is important to perform the rotation of figure 1.7 (temporal section of the human brain) in ones minds eye. Rotation inverts the relationship of the white matter and gray matter exhibited at the cortical epithelial margin (the cortical limbus). In the temporal lobe the white matter mechanism (i.e. the the fimbria-fornix) is positioned *on top* of the hippocampal formation. Invert this, and the white matter cables (fimbria-fornix, joined by the corpus callosum) are below the hippocampal formation. This is exactly what is encountered in a canal section through the body of the corpus callosum (i.e. the hippocampal formation (remnants) are positioned on top of the corpus callosum). And outside the vestigial hippocampal formation is the cingulate mesocortex.

We can therefore rotate figure 1.7A around the axis (figure 1.12A) and apply the relationships seen in figure 1.7A throughout the hemisphere. Notably, in the human brain the subdivision of the hippocampus in close proximity to the corpus callosum takes on a rudimentary appearance (losses its morphological differentiation). What this means is that by simply rotating figure 1.7 around the transverse hemispheric axis in the human brain, you do not *see* the expected anatomical relationships—in those hippocampal sectors close to the corpus callosum. It doesn't mean that they are not "there".[14] The same topologic relationships exhibited in figure 1.7 are

[14] A brief digression into comparative anatomy is helpful here. In the text it is stated that in most mammalian orders the subdivision of the hippocampus, which lies proximate to the corpus callosum involutes. However this is not the case in Chiroptera (the bat) nor in Edentata (a mammalian class containing the armadillo). Gloor points, that out that if you examine a hemispheric coronal section in bats that would correspond to a coronal section in the human brain passing through the body of the corpus callosum, you do see (in bats) "the unmistakable cytoarchitectonic pattern of hippocampus and dentate gyrus *above* the corpus callosum", see Gloor figure 2.23.

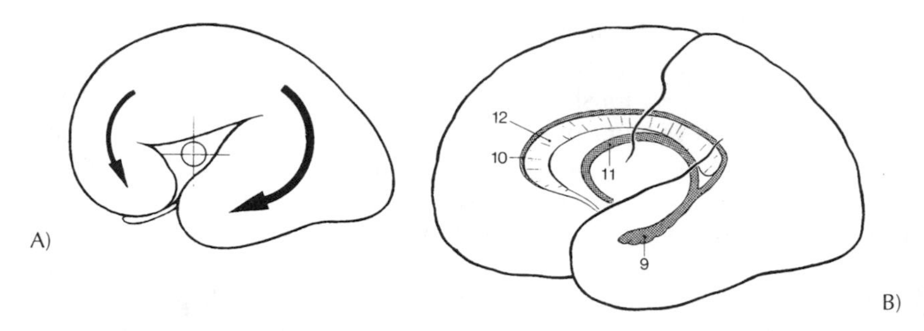

Figure 1.12. A) Spiral growth of the hemisphere around an axis drawn through the lateral fissure. This axis is denoted by the circle (target site) seen in the enlarged lateral fissure. The curved arrow demarcates the main hemispheric spiral. This spiral growth results in the elaboration of the temporal lobe in higher mammals. The limbic convolution is the gyral ring or arm proximate to the central axis of the spiral. B) Through the mechanism of the fornix, both spirals are joined. Figure 14, reproduced with permission from (3).

carried around the splenium although not evident in most mammalian orders.[15]

With this background we briefly overview hippocampal sectors throughout the hemisphere (excluding the head of the hippocampus).

Retrocommissural Hippocampus: Behind and below the splenium of the corpus callosum, the hippocampal formation is referred to as the retrocommissural hippocampus. In callosal mammals it is only in the retrocommissural district that the prototypical double-C architecture of the hippocampal formation is evident. In humans the double-C architecture is most evident in the temporal lobe section, figure 1.7. In the human brain, temporalization can be said to pull (rotate) most of the retrocommissural hippocampus away from the corpus callosum and into the temporal lobe, see figure 1.13 (Gloor figure 2.25, also see Gloor figure 2.35F). It should be appreciated that the retrocommissural hippocampus is divided into two sectors, the dorsal and ventral hippocampus (9, figure 2.35).

[15] Another problem is encountered with simply rotating the hippocampal formation around a transverse axis. The hippocampal formation would map out a circle (or near circle-like) pattern if simply rotated around the axis. Serial coronal sectioning of the hippocampal formation in mammals reveals, that the circular pattern is markedly kinked or deformed particular under the splenium of the corpus callosum. Indeed part of the hippocampal formation is folded back on itself under the splenium: this is referred to as the subcallosal flexure of the hippocampal formation. The subcallosal flexure is discussed later in the body of the text.

Subcallosal Flexure: On its approach to the corpus callosum, the hippocampal formation makes a sharp U-turn, as Gloor puts it, under the corpus callosum (figure 1.13). The U-turn is referred to as the subcallosal flexure, the retrocommissural hippocampus can be said to end at this flexure. The subcallosal flexure is situated at the midline underneath the corpus callosum, the dorsal aspect of the retrocommissural hippocampus approaches this midline flexure from its more lateral inferior position in temporal lobe: this explains the *appearance* of two hippocampal formations in coronal sections near the flexure, see Gloor, his figure 2.26). The appearance of both sections is a result of the more lateral inferior position of the dorsal hippocampus coming into conjunction with and the midline position of the subcallosal flexure. At the point of conjunction coronal section reveals the single non duplicated hippocampal formation.

Fasciola Cinera (*little gray gyrus*): On its passage from the subcallosal flexure, and the underside of the corpus callosum, around the splenium of the corpus callosum, to the *dorsal* surface of the corpus callosum, the hippocampal tissue is referred to as the fasciola cinera. The histological characteristics of the hippocampal formation are lost in the fasciola cinera.

Indusium Griseum: The fasciola cinera is interposed between the subcallosal flexure and the indusium griseum, the hippocampal tissue on the dorsal surface of the corpus callosum. The indusium griseum in most mammals is reduced to a thin poorly developed gray structure its cytoarchitectonic and topologic characteristics lost. However, as noted above (foot note 14), the bat is an exception, and a very helpful exception. In this species careful section reveals the cyotarchitectonic pattern of the hippocampus and dentate gyrus to be maintained even in areas proximate (just dorsal) to the corpus callosum.

Precommissural Hippocampus: Precommissural hippocampal tissue swings around the genu of the corpus callosum, and in higher primates continues a spiral like course in the direction of the anterior commissure en route to the base of the brain. The precommissural hippocampal tissue is the rostral most district of the hippocampal allocortex. It is also referred to as the taenia tecta (hidden band). The precommissural hippocampus is disposed around the white matter aspect of the border (around the fornix, septum pellucidum) in a manner similar to the gray white disposition of the cortical limbus in the temporal lobe, see figure 1.7A.

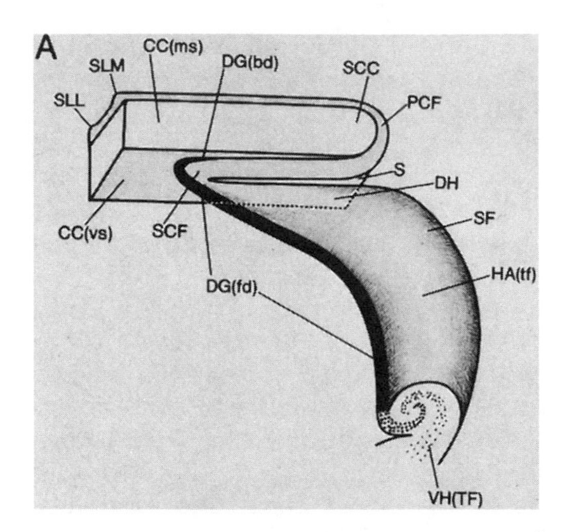

Figure 1.13. A midsagittal section through the hippocampal/limbic district. This diagram is unusually helpful because it embeds a cross sectional representation of the hippocampal tissue in proximity to the corpus callosum. The diagram also illustrates the behavior of the hippocampal tissue in proximity to the corpus callosum. The nature of diminutive hippocampal tissue opposed to both the undersurface and superior surface of the corpus callosum—a vexatious area indeed in neuroanatomy—is rendered less problematic by examination of this diagram. Reprinted with permission, from Gloor (9, p. 71).

WHITE MATTER SIDE OF THE HIPPOCAMPAL REGION: ALVEUS TO THE GREAT MIDLINE HEMISPHERIC COMMISSURES

In the previous sections of this chapter, the focus of our examination was primarily on the gray matter: a survey along the continuous line of cortex from the rhinal sulcus to the fimbrio-dentate suclus. This focus was intentional: in keeping with the strategies laid out at the beginning of the survey. In this section we shift our focus to the white matter side of the hippocampal, limbic border region.

Alveus to Fimbria

Fibers originating throughout the hippocampus assemble in a layer of white matter at the ventricular surface of the hippocampus. This lining of white matter is referred to as the alveus (L. belly). Nauta notes this white matter layer must have resembled a whitish belly (perhaps of a fish) to an earlier age of anatomists. The alveus similar to the deep fundal white matter of gyri throughout the hemisphere is found in the expected location: en face to the ventricle (although a roll of cortical tissue, the hippocampus proper is *not* referred to as a gyrus (in a way it is like a gyrus: gyrus in Latin means roll).

Having assembled in the alveus, the fibers of the hippocampus pass to the medial extension of the alveus, referred to as the fimbria fornicis (L. *fimbria*, fringe). The rather extensive whirlpooling of the hippocampal dentate continuum everts the white matter to such a degree that the fimbria faces toward the interhemispheric fissure and also toward the fornix and the midline cabling mechanisms of the hemisphere, see figure 1.7A.

Great Commissures of the Midline: (Fimbria-Fronix, Corpus Callosum, Anterior Commissure)

The fimbria-fornix[16] projection system brings into the ambit of discussion the midline commissural system. In this section we introduce some basic anatomic and embryologic principles pertaining to this commissural system. Admittedly, the discussion in this section is very broadly drawn.

It is generally agreed that all the major midline commissures grow out of a thickened area of the lamina terminalis: the thickened area is located in front of the interventricular foramina. This thickening is evident in the human by the fourth months of development (figure 1.14). The inferior aspect of this thickening, which will become the anterior commissure can be said to become constricted off during development: The anterior commissure appears positioned at the anterior inferior extreme of an imagined inner hemispheric spiral mapped out by the midline commissures (one can backproject a connection between the anterior commissure and the more dorsally positioned commissural mechanisms). The upper more rostrodorsal part of the thickened lamina terminalis grows with the expanding hemisphere. Transverse fibers passing between the isocortical districts of the hemisphere pass into the most dorsal aspect, which becomes differentiated as the corpus callosum. Longitudinally directed fibers of hippocampal origin passing between the lamina terminalis and posterior hippocampal districts enter in the ventral part, and thus the ventral part becomes differentiated as the fornix. A thin walled portion of the lamina terminalis situated between the corpus callosum above and fornix below receives no fibers, and evidences a pellucid appearance. This portion is referred to as the septum

[16] The fimbria is continuous with the fornix. The change in terminology occurs at the splenium. Fornix is probably employed because the term reflects the arc-wise nature of the pathway, whereas fimbria refers more specifically to the gathering fibers from the alveus.

pellucidum: the septum pellucidum can be interpreted as a stretched out portion of this midline system.

Examination of the behavior of the callosal system, in phyloge-netic sequence[17], provides an additional perspective on the midline cabling system. In acallosal forms, such as the marsupial, there are but two commissures: the large anterior commissure, and a smaller more dorsally positioned hippocampal commissure (both commissures serving to connect the allocortical districts). In the marsupial, the hippocampal tissue lies undistorted in a crescent outside (mainly dorsal) to the anterior commissure. "Ascending" the phylogenetic scale, a new commissure appears, the corpus callosum, which can be seen developing in a position yet more dorsal to the 'older' anterior, and hippocampal commissures, see figure 1.15.

DEVELOPMENT OF THE CORPUS CALLOSUM

Our treatment of the midline commissural system in the preceding section was a summary overview which was correct in concept, but lacking in detail. Since a principle concern of this chapter is an examination of the corticoepithelial border, it would be inconsistent to leave the description at this point. However, to continue the discussion in a proper fashion would require a more extended narrative, and the use of a rather extensive iconography (fortunately available in such recent texts as Gloor's 9, pp. 67–81). For the purpose of this text, we will address only one of the issues facing investigators of corpus callosal development throughout most of the last century (an area of study made more problematic because of the absence of transitional forms: recall there are acallosal mammals, and callosal mammals, and no in intermediate forms). The issue was, where were the initial crossing callosal fibers laid down? Everyone agreed that the initial crossing occurred in the dorsal extremity of the lamina terminalis (as depicted in standard cross sections of the lamina terminalis); however it was not clear exactly where. For example, Elliot Smith suggested that crossing callosal fibers split apart the anlage of the hippocampus in such a way that the constortium of structures constituting the hippocampal formation would ride on the back of the corpus callosum (the crossing white matter fibers positioned themselves underneath the hippocampal formation).

[17] There are three major groups of mammals. All placental mammals have a corpus callosum. The other two groups of mammals (two orders: Marsupialia, and Monotremata) are acallosal. There are no transitional forms. See Gloor's text for a brief overview of theories on the origin of the corpus callosum.

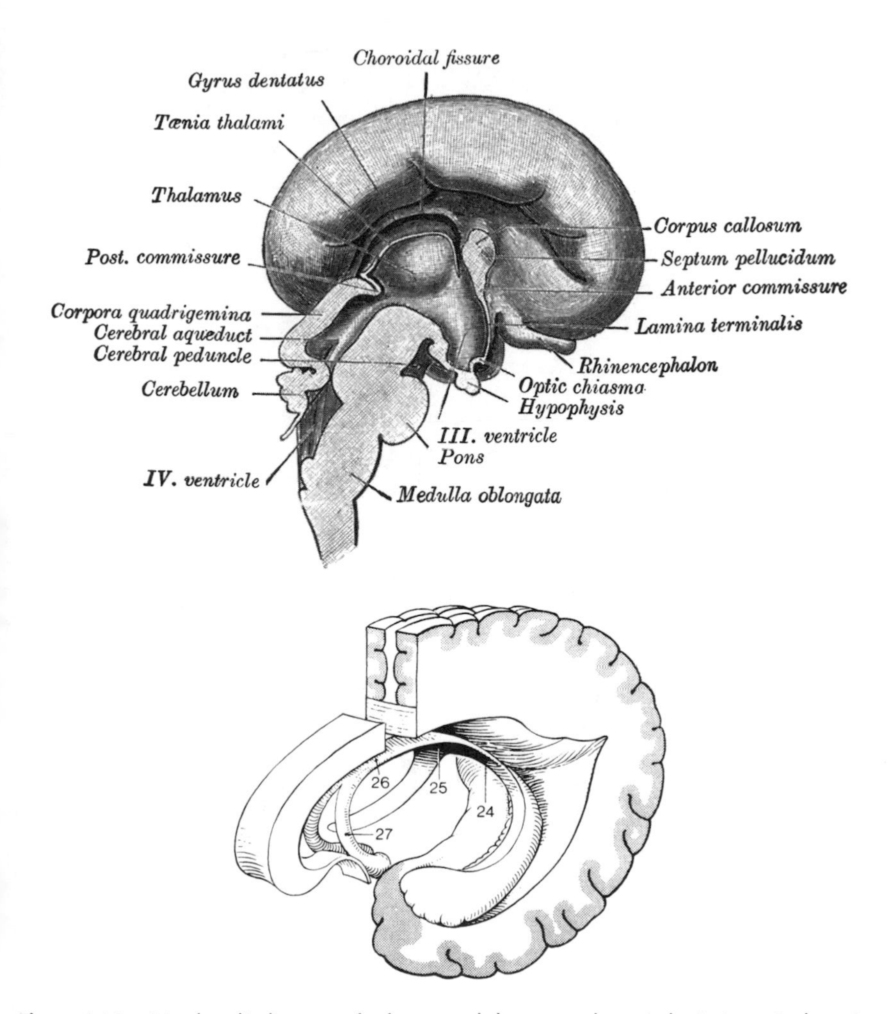

Figure 1.14. Marchand's diagram of a four month human embryonic brain in sagittal section demonstrates that the great commissures of the hemisphere all initially appear as a thickening of the lamina terminalis in front of the interventricular foramen. As the hemisphere grows, the commissural white matter tracts can be said to spiral out of this thickening, a phylogenetic overview of commissural development is discussed in the chapter. Figure from Gray's Anatomy 24th edition (15).

Other investigators (Abbie), for example, argued for a more dorsal passage of the early callosal fibers: a more dorsal position would result in the callosal apparatus carrying the hippocampal formation on its belly. A century of investigative efforts has resolved this issue, and the resolution has been

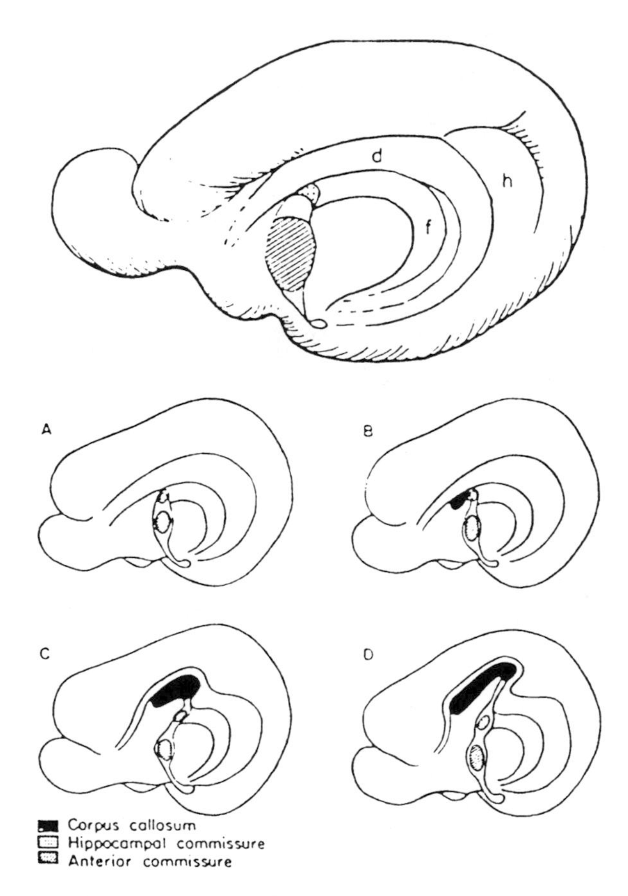

Figure 1.15. A) Medial view of the acallosal marsupial hemisphere. B) Superimposed on the marsupial plan is the commissural arrangement in the more advanced callosal mammal. Note how the corpus callosum develops dorsal to the older commissures. Reprinted with permission, from White (13, p. 9).

made available at the level of textbook anatomy (Gloor, Kuhlenbeck). In this text, examination of figure 1.7A accords with Elliot Smith's argument. Flip 1.7A and the hippocampal formation does seem to ride on the back of the fimbrial fornix system throughout its non temporal lobe course. Examination of figure 1.13 (Tilney's analysis, as presented in Gloor) sheds additional light on how the hippocampal formation can ride both on top and underneath the callosal white.

Extended Definition of the Hippocampal Formation

Although we have only touched on the complexities of corpus callosal development, the close continuity exhibited by the earliest crossing fibers of the corpus callosum, and the epithelial area of the hippocampal region (the epithelial area being the seat for the main projection area of the hippocampal formation) underscore the arbitrary nature of restricting the hippocampal formation to just the hippocampus proper and the dentate. In this regard an extension of the term is offered to include the white matter mechanisms (9, p. 79).

More importantly, a study of callosal embryogenesis underscores the invariance of gray white relationships, the preservation of topology. Gloor's comment closes this section.

> "The continuity of the fimbria-fornix and the concept of regarding the corpus callosum . . . as part of the hippocampal formation make it clear that the original topology of the hippocampus and of the surrounding structures has been preserved in the brain of callosal mammals."

BROCA'S LIMBIC LOBE: AN EXAMPLE OF THE SEARCH FOR HEMISPHERIC ARCHETYPE

"A universal integrity of spirals unites all creation."

SCHNEIDER

"The idea of subdividing the mesial suface of the hemisphere into a series of concentric areas seems to exert some strange fascination over the minds of morphologists within recent years, thanks largely to the writings of Zuckerandl. What exactly is gained when the mesial surface has been thus arbitrarily split up into "Banwindungen is not very evident, unless it be the fact that Broca's ideas of a "limbic lobe and Schmidt's conception of "Bogenwindungen" both find expression in this exercise of the imagination, generally so futile and meaningless."

ELLIOT SMITH

"Upon supposed morphologcal grounds, Broca separated the olfactory bulb and its peduncle, the locus perforatus anticus, the uncinate gyrus, and the callosal gyrus as a great cortical ring completely surrounding the hilum of the he'misphere. He supposed these regions to be still further united physiologically in the sense of smell.

But to such a subdivision there Is an insurmountable objection"

ELLIOT SMITH (21, p. 444)

In the introduction to this chapter, it was argued that Broca's proposal of a great limbic lobe represented an attempt by neuroanatomists in the 19th century to identify the architectural principles underlying mammalian hemispheric design, a search for unity in diversity (a search for the morphotype (see the sidebar on page 11–12, also Table 2, also figues 1.2, 1.4, and 1.5). Although not all neuroanatomists ascribed to the value of such Platonic visions (see Elliot Smith's comments given in the epigraphs above).

The marginal band of tissue at the hilar area of the mammalian brain, which has entrained our attention throughout most of this chapter, does seem to map out, an underlying spiral like geometry to the temporal lobe, see figures 1.7A (coronal section), and also to the main hemispheric convexity, see figure 1.3 (sagittal section). These spirals stand at right angles to each other, see figure 1.12. Again, with our attention drawn to the limbic region, particularly by the fimbria-fornix, and stria terminalis, the spirals seem to come into conjunction. Indeed, it can be argued that through such intermediaries as the fornix, and stria terminalis, the central axis of the temporal and main hemispheric spirals are conjoined. Of course, the validity of such hemispheric design principles suggested by gross anatomic inspection of the adult form, only become fully intelligible in the context of comparative anatomy, also (and not just a geometric idle/exercise).

Notably, Gloor de-emphasizes the idea of an imagined spiral growth around an axis positioned through the lateral fissure. The 'imagined rotation' as he puts it, is better understood to result from the caudal expansion of the isocortical areas and the associated expansion of the thalamus and internal capsule in relationship to the fixed anchor points of the vertebrate hemisphere.

A TABULAR VIEW OF THE LIMBIC BRAIN CONCEPT

The last three sections of this chapter and the sidebar on this (the following) page are presented to buttress basic concepts of limbic brain anatomy. These sections provide as wider compass of introductory description, helpful to the reader new to this area.

The Cerebrovascular Hilum

"The limbic system (MacLean, 1952) has its origin in the name 'grande lobe limbique' coined a century ago by Broca (1878) to denote the somewhat distinct region of the cortex nearest the margin (limbus) of the cortical mantle,

a region which in the mammalian brain fully encircles the hilus of the cerebral hemisphere."

(27, p. 175)

Conceptualizing the cerebral hemisphere as built up out of a set of concentric rings enables an appreciation of an important issue in cerebrovascular anatomy. During organogenesis, the brain, similar to the lung or kidney, or the head of a mushroom, can be envisioned as growing out from and around a central axis. The plane of this axis remains relatively stationary receiving little mechanical displacement during development. In this plane is seated the vascular portal or hilum of the brain, (i.e., the circle of Willis). The vascular portal is nested just within the limbic crescent.

The penetrant vessels which branch off the circle and innervate the deep gray parenchyma of the hemisphere core (lenticulostriate, and thalamoperforants) are of relative thin caliber (paticularly in relationship to the circle vessels) and exhibit sensitivity to hypertensive changes (undergoing a pathophysiologic process termed, lipohyalinosis). In the setting of hypertension, such changes set the stage for the development of disseminated infarcts (multi-infarct dementia) also frank cerebral hemorrhages in the deep gray striatal crescent (also capsular area) surrounding the circle of Willis. Importantly, these small vessels feed the limbic system and associated subcortical sites. The impact of small vessel disease in the perihilar (limbic) area of the brain brings about, not unexpectedly, profound neurobehavioral changes.

In this section, we introduce Mesulam's iconography of all major cortical neuronal types. Mesulam's table provides a different perspective (a tabular iconography if you will) on basic concepts of limbic brain anatomy. It should be appreciated that Mesulam's table was not designed to depict a cytoarchitectural cortical tapestry, nor to address basic concepts of the limbic lobe. Mesulam's table was from his chapter introducing principles of behavioral anatomic correlation.

Only the following main points more immediate to limbic brain anatomy suggested by the table are outlined below.

— In his table, adapted here, see Table 3, Mesulam divides the entire hemispheric surface into five basic types of cortex (notably, the division of the cortex into five major types did not originate with this table). The five cortical types range from the most simple, poorly laminated cortical like (i.e. corticoid) tissue found in the such forebrain locations as the septal area, and amygdaloid body (although the amygdala is largely a subcortical structure, Mesulam considers it corticoid

Table 1.3. A Tabular Iconography of the Cerebral Cortex

EXTRAPERSONAL SPACE
Exteroceptive Function
Primary Sensory and Motor Cortex
Association Cortex (Unimodal)
Association Cortex (Heteromodal)
Paralimbic Areas
Temporal pole—caudal orbitofrontal
Anterior insula—cingulate—parahippocampal
Limbic Areas: Septal substantia inominata—amygdala
piriform cortex—hippocampus (Corticoid + Allocortex)
Hypothalamus
Interoceptive Functions
Internal Milieu

See Text for Details. Adapted, with permission, from (26).

or cortical like because of its well recognized role in complex associative, both emotional and cognitive, processes) to the most differentiated, six layered isocortex constituting the primate sensory and motor and association areas.

— Note how the lower two tiers of Mesulam's table, the limbic, and paralimbic areas respectively, cohere with a description of the limbic brain in terms of the growth ring hypothesis. Mesulam's inner most tier is constituted by allocortex; the neighboring tier named the paralimbic areas is constituted by transitional or mesocortical cortex.

— The inherent robustness of core limbic connections within the hypothalamic[18]-limbic-paralimbic axis (Nauta's extended hypothalamus) is readily evident even from a cursory glance at the table. Hypothalamic connections to the limbic and paralimbic tiers was a central element in Papez's circuit, see chapter 2. Paralimbic connections to the associational cortex (intimated at by Papez) received particular emphasis in the constructs of MacLean and Nauta, see chapters 3 and 4.

— Mesulam's tabular iconography depicting cortical limbic connections can be easily whirlpooled into a circular iconography, contrast table 3 with figure 1.5.

NAUTA'S SEPTO-HYPOTHALAMIC CONTINUUM: INTRODUCTION TO A LINEAR LIMBIC ICONGRAPHY

"A second conceptual realm of the forebrain is constituted by the limbic system, a collectively functionally neutral term denoting a heterogeneous

[18] The term hypothalamus was not known to Broca: it apparently originated with Edinger in the late 19th century, (7, p. 248). Broca's did include his carrefour de l'hemisphere, also his diagonal band in his work, both of these structures (his term carrefour de l'hemisphere is now just an historical curio) are located in the basal forebrain proximate to hypothalamic parenchyma.

group of neural structures arranged along the medial edge of the cerebral hemisphere and having at its core the allocortical hippocampus and the largely subcortical amygdala. Together these from the principal telencephalic origin of the medial forebrain bundle, and thus, the hemisphere's main link with a continuum of interconnected subcortical structures that begins in the septum and extends caudalward from there through the hypothalamus and midbrain to upper levels of the rhombencephalon."

(2, p. 43)

A particularly straightforward introduction to the concept of the limbic brain was provided by Nauta. Nauta directed out attention to the idea that the limbic brain most simply put could be conceived as a neural continuum (consisting of subcortical and cortical elements) in which the hypothalamus is the central element (16, 27). The principle components of this continuum (a linear iconograph) centered on the hypothalamus are the septum, and paraseptal area, the amygdala, and the gyrus fornicatus (notably these components are grouped in the lower two tiers of Mesulam's table).

Experimental and behavioral data in support of this grouping include the fact that many if not all of the effects produced by stimulation and lesion of the extra hypothalamic limbic structures can be replicated by stimulation of lesion of the hypothalamus. Furthermore, these core limbic loci (hypothalamus, the amygdala, also the hippocampus) although not proximate to each other in the "higher" mammalian brain, were indeed in close proximity (nested in the medial hemispheric wall) in the ancestral form, as evidenced today by the amphibian brain.

SUMMARY

In this chapter, we have overviewed classical (19th century) description of the limbic brain. We then explored the limbic, hippocampal area using more modern illustrations, and current terminology (our exploration for the most part focused on the body of the hippocamus, in the temporal lobe). Finally, we surveyed a variety of introductory approaches to the topic of limbic brain anatomy.

Broca who helped usher in the great age of brain behavior functional correlation in his seminal work in aphasia would hardly be one to have been reticent regarding a behavioral role for his limbic lobe. Indeed he ascribed a principal olfactory function to his limbic lobe (recall that the anterior aspect of his limbic lobe was constituted by the olfactory bulb and tract). However, it was clear to Broca that the limbic lobe must be contributing to an alternative, not purely olfactory, function (Broca knew for example that mammals devoid of an olfactory sense (i.e., Delphinidae) still possessed significant anterior cingulate tissue. Broca, as we shall see in chapter 2, did intimate at an alternative limbic function: an emotive function, the limbic lobe as the mediator of the l'homme brutale. However with no experimental data or case studies to develop this line of thought, Broca's suggestion was not developed.

Three quarters of a century later, based on accumulated empirical data (clinical findings, data from lesion experiments) and theoretical developments regarding the division of labor between the hemispherical walls, James Papez would see in Broca's circumannular convolution a mechanism for the anatomical elaboration of emotion. We turn to Papez's celebrated circuit in the following chapter.

REFERENCES

1) Broca, P. (1878). Anatomie comparee des circonvolutions cerebrales. Le grand lobe limbique et la scissure limbique dans la serie des mamiferes, *Rev. Anthropol. 1*, Ser. 2, pp. 385–455.
2) Nauta, W. (1986). Circuitous connections linking cerebral cortex, limbic system, and corpus striatum, in *The Limbic System and Clinical Disorders* (B. K. Doane, and K. E. Livingston, eds.), Raven Press, New York, pp. 43–54.
3) Kahle, Leonhardt, and Platzer (1993). *Color Atlas of Human Anatomy, V. 3. Nervous System and Sensory Organs*, Thieme Medical.
4) Lewy, F. H. (1942). Historical introduction: The basal ganglia and their diseases, in *The Disease of the Basal Ganglia*, Res. Publ. Assoc. Res. Nerv. Ment. Dis. 21, p. 1–20.
5) MacLean, P. D. (1990). *The Triune Brain in Evolution*, Plenum, New York.
6) Karten, H. (1969). The organization of the avian telencephalon and some speculations on the phylogeny of the amniote telencephalon, *Ann. N.Y. Acad. Sci.* 167, 164–179.
7) Schiller, F. (1979/1992). Paul Broca, Explorer of the Brain, Oxford University Press.
8) Smith, G. E. (1902). On the homologies of the cerebral sulci, *J. Anat.* 36, 309–319.
9) Gloor, P. (1997). The Temporal Lobe and Limbic System. Oxford University Press.
10) Duvernoy, H. M. (1991). The Human Brain, Springer-Verlag.
11) Smith, G. E. (1896). The morphology of the true "limbic lobe" corpus callosum, septum pellucidum and fornix. *J. Anat. Physiol.* 30, 157–167.
12) Rolando, L. (1830). "Della Strutture delgi Emisferi Cerebrali," Mem. R. Accad. Sci Torino, Vol XXXV, pp. 103–146, from Schiller, F., p. 257, 1992.
13) White, L. E. (1965). A morphological concept of the limbic lobe. *Int. Rev. Neurobiol.* 8, 1–34.
14) Duvernoy, H. M. (1998). *The Human Hippocampus*, Springer-Verlag.
15) Gray, H. (1942). *Anatomy of the Human Brain* (W. H. Lewis, ed.), 24th Edition, Lea, and Febiger, Philadelphia.
16) Nauta, W. J. H., and Fiertag, M. (1986). *Fundamental Neuroanatomy*, W. H. Freeman, and Company, New York.
17) Angevine, J. B. (1975). Development of the hippocampal region, in *The Hippocampus Volume 1: Structure and Development* (R. L. Issacson, and K. H. Pribram, eds.), Plenum Press, New York, pp. 9–39.
18) Yakovlev, P. I., Angevine, J., and Locke, S. (1966). The limbus of the hemisphere, limbic

nuclei, of the thalamus, and the cingulum bundle, In: *The Thalamus*: D. Purpura, and M. Yahr. Editors. Columbia University Press, 77–97.
19) Meyer, A. (1971). *Historical Aspects of Cerebral Anatomy*, Oxford University Press.
20) Lewis, F. T. (1923–24). The significance of the term hippocampus. *J. Comp. Neurol.* 35, 213–230.
21) Smith, G. E. (1901). Notes upon the natural subdivision of the cerebral hemisphere, *J. Anat. Physiol.* 35, 431–454.
22) Northcutt, R. G. (1969). Discussion of the Preceding paper. Ann N. Y. Acad Sci.
23) Kuhlenbeck, H. (1973). *The Central Nervous System of Vertebrates: Overall Morphologic Pattern.* V. 3, Part II S. Karger.
24) Butler, A. B. (1994). The evolution of the dorsal pallium in the telencephalon of amniotes: cladistic analysis an a new hypothesis, Brain Research Reviews 19, 66–101.
25) Isaacson, R. L. (1982). The Limbic System, Plenum Press, New York.
26) Mesulam, M.-M., (1985). Principles of Behavioral Neurology, F. A. Davis, Philadelphia.
27) Nauta, W. J. H., and Domesick, V. B. (1982) Neural Associates of the Limbic System, in *The Neural Basis of Behavior* (A. L. Beckam, ed.), pp. 175–206.

CHAPTER TWO

PAPEZ'S CIRCUIT

Kingsley spoke slowly,
"As far as I am aware, these events can be explained very simply, on one hypothesis, but I warn you it's an entirely preposterous hypothesis"

FRED HOYLE, The Black Cloud

". . . it is proposed that the hypothalamus, the anterior thalamic nuclei, the gyrus cinguli, the hippocampus and their interconnections constitute a harmonious mechanism which may elaborate the functions of central emotion, as well as participate in emotional expression. . . ."

1) JAMES PAPEZ, p. 743, 1937

"I must say here that, after having read (probably early in 1950) Paul MacLean's paper of 1949 on the 'visceral' brain, which lead me to Papez's paper of 1937 on the 'mechanism of emotion', I was excited by the feeling of not being alone in my search for a definable and measurable 'substance' for the concept of human 'psyche' (at the peak of Freudian dispersion in those years), . . . I was intrigued by the din and babble on 'emotions' in psychiatry in the 1930's and 1940's, but didn't quite know how to deal with 'emotion' myself without provoking virulent emotions in some of my otherwise genial friends at that time. . . . From the impact and shock of contradiction new truths will eventually emerge . . ."

2) P. YAKOVLEV, (1978)

INTRODUCTION TO PAPEZ'S CIRCUIT

"I want to say a few things about each part . . . There are grossly speaking, two kinds of nervous system organizations . . . The first is the brain stem, together with the limbic (hedonic) system, the system concerned

> with appetite, sexual and consummatory behavior . . . It is a value system . . ."
>
> 3) Gerald Edelman, p. 117, 1992

> "These 'values' drives, instincts, intentionalities—serve to differentially weight experience to orient the organism toward survival and adaptation, to allow what Edelman calls, 'categorization on value' . . . in the words of the late philosopher, Hans Jonas, "the capacity for feeling, which arose in all organisms, is the mother-value of all."
>
> 4) Oliver Sacks, 1993

In 1937, James. W. Papez (1883–1958) published a paper entitled, A Proposed Mechanism of Emotion (1), although Papez's paper initially received limited attention, it would go on to create, in the words of Schiller "quite a stir: a significant step along the wild goose chase" opened by Gall at 'localizing' in the brain a psychological function: emotion (5, p. 248). With a synthetic vision reminiscent of Kepler (6), Papez's idea, at its most simplest, proposed that a circularly arrayed consortium of neural components and their connections participated as an harmonious mechanism (a mechanism known to us now as Papez's circuit) elaborating emotional awareness. In the wake of Papez's conjecture, emotional consciousness would be no more or less mysterious, and would have anatomic parallels to other conscious/aware senses, e.g, the somesthetic senses (i.e., touch pain, joint) also vision.

The major elements composing Papez's neural circuitry conformed in large measure to Broca's great limbic lobe (7). However, the similarity of Papez's circuit, to Broca's great limbic lobe received only an oblique and passing mention by Papez (1, p. 734). This was hardly a slight of his predecessor. Broca's speculations regarding an emotive function for his limbic lobe, even his allusion to his limbic lobe as the cerveau brutale simply did not contribute in any *direct* way to the development of Papez's concept: this could not be otherwise. Basic concepts of hemispheric wall function—e.g., relating the medial wall to a visceral/emotive role; and the lateral hemispheric wall to the governance of somatic/body wall function (well understood by Papez, and a principal aspect of his theory)—were essentially unknown to Broca. (However, Broca did suggest a role for his great limbic lobe that did, in a way, foreshadow the work of C. J. Herrick, Elliot Smith, and Papez, see the sidebar).

DID BROCA SUGGEST A VISCERAL OR EMOTIONAL ROLE FOR HIS LIMBIC LOBE?

AN ENERGIZING, REINFORCING (REWARD?) FUNCTION

"The affective role of visual sensations, though arguably present, is not nearly so striking as with smells"

NICHOLAS HUMPHREY, A History of the Mind, 1992

Although Broca disclaimed any certain knowledge of a particular behavioral role for his great limbic lobe, stating early in his paper, *"The name limbic convolution that I have adopted indicates the constant relationship of this convolution with the border of the hemisphere; it does not imply any theory . . ."* (7 p. 394), he was not silent regarding possible behavioral (emotional) correlations. Broca did favor the idea that the limbic lobe was principally an olfactory organ. However, aware that several mammalian forms were microsmatic, such as the dolphin, and human; yet had well developed anterior cingular and hippocampal convolutions, Broca advanced the idea that his limbic lobe might play a role in olfactory ***ASSOCIATION***, and in the context of this associational role the limbic lobe might function as a ***MOTIVATIONAL*** agency, *"L'odorat joue chez eux un role . . . dans le choix de la nourriture, dans la poursuite de la proie, dans la fuite du danger, dans la recherche de la femelle, dans le retour au gite"*. (7, p. 393). Broca seems to be implying that the animal is apperceiving olfactory sensation from the animal's point of view with respect to pleasure or pain, see Schiller (5) for further discussion of this issue. A half century after Broca, Herrick pursued exactly this line of thought when he proposed a forebrain regulator of affect, based on an evolving nexus of cross associational pathways strongly linked to olfactory centers, (8, p. 13).

LE CERVEAU, AND LE CERVEAU INTELLECTUEL:

Broca also expressed the opinion that the limbic lobe functioned more as an animalistic ("brutale") brain, an imperfectible part of the hemisphere, the seat of inferior faculties. The rest of the cerebral mantle he referred to as the intellectual brain—the seat of superior faculties, the intelligent animal (7, p. 420). Broca's ideas regarding behavioral functions for his limbic lobe were rather sketchy and did not receive any following in the neuroscience community of his time. For example it should be noted that Broca was not alone amongst nineteenth century anatomist, in the construction of bipartite anatomic and behavioral divisions of the hemisphere. Franz Joseph Gall (1758–1828) suggested a back to front ordering parameter for behavior (with posterior elements and the cerebellum subserving more lower faculties);

Theodore Meynert (1833–1892) also offered bipartite views on hemispheric architecture: Meynert argued for a physiological antagonism between subcortical and cortical structures (attributing psychotic disturbance to subcortical disturbance) and Meynert also promoted the idea of a motor-anterior versus sensory-posterior dichotomy in hemispheric function. Each of Meynert's views influenced respectively two of his more famous pupils, Freud, contributing to the development of his theory of psychosis, and Wernicke, contributing to his proposal of sensory posterior aphasia in contrast to anterior motor aphasia (9,10,11).

The rather sudden appearance of Papez's paper on the mechanism of emotion (appearing as it did all at once, as a rather sophisticated anatomic hypothesis, incorporating behavioral medicine, neuropathology, comparative anatomy, and evolutionary theory) begs for an examination of developments in the neuroscience that set the stage for its birth. If we exclude Broca as a conceptual forebear of Papez's proposal, where can one trace the foundations to Papez's circuit? A detailed analysis of the historical antecedents to Papez's formulation is beyond the scope of this chapter. We will restrict our efforts to a brief discussion of four areas, which provided the underpinnings of Papez's circuit proposal. These areas included developments in brain transection and stimulation; developments in comparative neuroanatomy; the accumulation of clinical data; also an evolving neuroanatomic philosophy, which implicated reciprocating cortical to subcortical circuits as central processes elaborating consciousness.

BRAIN STIMULATION AND TRANSECTION

Brain transection work has a rich heritage dating to the early nineteenth century with the work of Flourens. However, the role of brain transection, also brain stimulation in pointing to central (rostral brain stem, and diencephalon) controlling elements in emotional behavior would not be realized until the late nineteenth century, and very early twentieth century with the work of Franz Goltz, Vladimir Bekhterev, and Karpus and Kried (12).[1]

Brain Transection: Pseudaffective Reflex State

It is in the brain transection experiment of Richard Woodworth (1869–1962) and Charles Sherrington (1857–1952), in the opening decade

[1] For a brief review of the *short comings* of lesions studies see (13,14).

of this century, that one first clearly recognizes the terminology and basic concept that would contribute directly to Papez's formulation. In 1904, these authors described what they referred to as a pseudaffective reflexive state observed in decerebrate cats (hemispheres including the diencephalon removed). The authors described the behavioral phenomena evidenced by these animals as follows; "*The truncation of the brain of the mammal at the mesencephalon annihilates the neural mechanism to which the affective psychosis is adjunct. But it leaves fairly intact the reflex motor machinery whose concurrent action is habitually taken as an outward expression of an inward feeling.*" (15, p. 234). Because these behaviors occurred in decorticate animals, and seemed to be devoid of true emotionality, they were designated, 'pseudaffective reflexes'. This experiment clearly implicated the upper brainstem in emotional expression: a finding central to Papez's theory.

Brain Transection: Sham Rage

In a series of transection experiments undertaken to better characterize the central regulation of emotional behaviors and sympathetic activity, the physiologist, Walter Cannon (1871–1945) described the behavioral phenomenal of sham rage. This behavior (similar to that observed in decerebrate dogs by Goltz) was in the words of the authors, "*a group of remarkable activities such as are usually associated with emotional excitement*..." (16, p. 285). Because the animals they observed had no cortical tissue, the authors felt that the activity could only represent the outward appearance of emotion, "a sort of sham rage" as they would put it. Widespread sympathetic activity was a feature of this state (evidenced by spitting, clawing, arching of the back, tail-lashing; also elevations of adrenalin output by the medulla). This feature pointed to rostral brain stem, and diencephalic integration with the peripheral autonomic nervous system. The state of sham rage differs from normal anger in that it can be elicited by very mild stimuli; additionally it sometimes occurred spontaneously in the prepared animals.

Philip Bard (1898–1977), working in Cannon's laboratory was determined to further characterize central components mediating such states as sham rage and pseudaffective behaviors. Bard was not satisfied that previous experimental efforts clearly demonstrated what part of the caudal diencephalon must remain intact (connected with lower centers) to maintain sham rage. In a series of experiments designed to answer this question, Bard performed serial transections on the neuraxis of the cat brain. Bard demonstrated that sham rage was attenuated as transections progressed to more caudal parts of the diencephalon, see figure 2.1. Bard's finding which

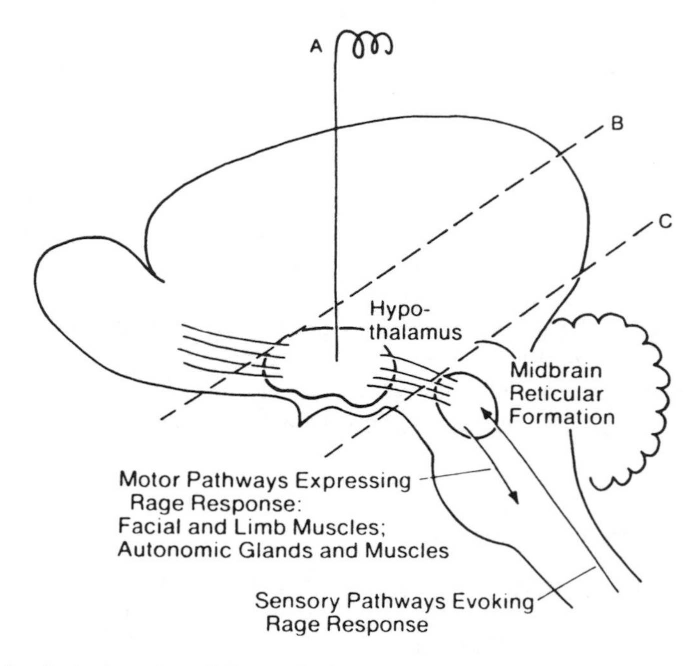

Figure 2.1. Sagittal section of the cat brain. The dotted lines represent brain transections used to study sham rage. Transection at level (B) leaving the hypothalmus intact is associated with sham rage. Transection at (C) removing the body, and most of the caudal aspect of the hypothalamus, abolishes sham rage leaving only disarticulated fragments of the rage response. Stimulation of the electrode (A) implanted in the hypothalamus produces integrated expressions of rage and fear. Walter Hess's efforts in electrical stimulation (which won him a Nobel Prize) were conducted contemporaneously with Papez's efforts. Interestingly however, Papez did not reference electrical stimulation in his paper. Figure reprinted with permission from (17, p. 573).

pointed to a mediator of emotionality (emotional expression) in the diencephalon would be cited by Papez on the second page of his paper (see following sections).

CONTRIBUTIONS FROM COMPARATIVE NEUROANATOMY

Central to Papez's proposal were developments in comparative studies revealing a model of vertebral hemispheric anatomy and function in which the medial aspect of the hemisphere was felt to be more closely integrated with internal organ (visceral, autonomic nervous system) homeostasis; and the lateral aspect of the hemisphere was understood to be principally

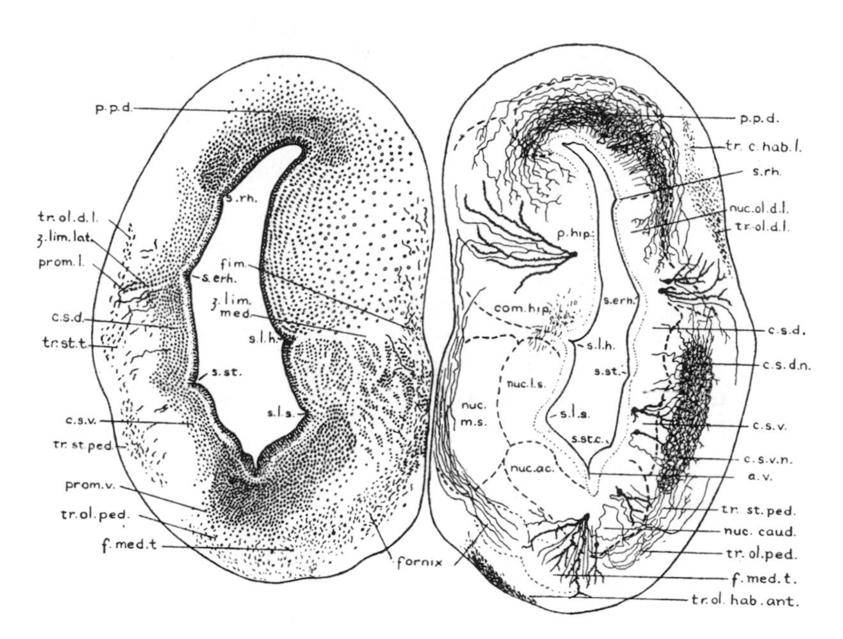

Figure 2.2. Transverse section of the tiger salamander's brain through the telencephalon. Comparative anatomical studies in the first half of this century suggested that the medial wall of the cerebral hemisphere bore a fundamental relationship with the hypothalamus, and the lateral wall of the hemisphere stood in relationship with dorsal thalamus (the term dorsal thalamus is employed here using Herrick's concept (19) of the "four floors" or four divisions to the diencephalon present in all vertebrates: the epithalamus, dorsal thalamus, ventral thalamus, and hypothalamus. The amphibian brain is lissencephalic; cingulate gyral cells are not evident; the primordial hippocampi is connected with the hypothalamus in a very diffuse way—through a rudimentary fornix. The salamander's brain serves (because brains don't fossilize) as a representative of nature's plan for the hemisphere as it may have appeared in the mid to upper Paleozoic (~ 400 million years ago). The urodele Amblystoma (the tiger salamander) was one of Herrick's preferred research specimen representing the simplest vertebrate brain: providing the ancestral pattern, the basic blueprint. Terminology employed: P. Hip., primordium hippocampi; C.S.D., dorsal part of the corpus striatum; C.S.V., ventral part of the corpus striatum; Nuc. M.S., and Nuc. L.S., medial and lateral septal nuclei respectively; Fim. Fimbria. Figure from Herrick, reproduced with permission (20)

involved with the coordination of somatic activities.[2] Figure 2.2, a cross section of the tiger salamander brain, from Charles Judson Herrick's (1868–1960), classic book helps to illustrate the conceptual division drawn

[2] The terms visceral and somatic have an anatomic as well as a related phylogenetic description. *Phyletic View*: The vertebrate body appears to have evolved from an acorn worm like organism, which was constituted mostly by a pharyngeal feeding basket, a gut tube and its

between medial and lateral hemispheric wall anatomy. As evident in the following citation from the first page of his paper, this perspective was a critical element in Papez's proposal

> "It is generally recognized that in the brain of lower vertebrates the medially wall of the cerebral hemisphere is connected anatomically and integrated physiologically with the hypothalamus and that the lateral wall is similarly related to the dorsal thalamus (Herrick). These fundamental relations are not only retained but greatly elaborated in the mammalian brain by the further development of the hippocampal formation and the gyrus cinguli in the medial wall and of the general cortex in the lateral wall of each cerebral hemisphere."
>
> ... Step by step, the structures of the medial wall become differentiated into the hippocampal formation, which establishes the first efferent connections of the cortex, namely, the fornix, with the hypothalamus ... The histories of the two walls of the hemispheres owe their disparity and distinctive structure to two totally different kinds of integration—the hippocampus and the cingular cortex participating in hypothalamic activities and the lateral cortex in the general sensory activities mediated by the dorsal thalamus. It is also noteworthy that in both systems two way connections exist between the cortical and the thalamic level."
>
> 1) pp. 725–726.

THE PROXIMITY OF THE HYPOTHALAMUS TO THE AMYGDALA, AND HIPPOCAMPUS

As noted in chapter 1, the hypothalamus, the mainly subcortical amygdala, also the allocortical hippocampus are generally considered core limbic structures. However, in the higher mammalian hemisphere these structures are not in close proximity to each other, although they are united by a rich neural interface.

appendages (diverticuli from this tube, e.g. liver, pancreas, lungs). The pharyngeal basket and gut tube are referred to as the visceral components (18, Romer's figure 4). Somatic structures of the vertebrate body are layered around (over evolutionary time, added to) the pharyngeal basket. *Anatomic View*: Visceral organs are generally tubular in nature invested with one or several smooth-muscle coats. The anatomic relationships (the topography) of visceral and somatic organs in all vertebrates are evident in the body of the worm like ancestor.

Even a cursory examination of the amphibian hemisphere (which serves as a representative of the ancestral vertebrate hemisphere), see figure 2.2, suggests the close proximity of these neural structures that once prevailed in a long extinct ancestor. The pathways mapped by the stria terminalis, stria medullaris, also the fornix, maintain ancestral relationships and serve as maps, recapitulating evolutionary pathways.

SUMMARY TO PREVIOUS SECTIONS

> "... in light of these researchers [Papez is referring to such figures as Herrick, and Bard] the connections of the hypothalamus to the medial wall of the cerebral cortex gain a new significance."
>
> PAPEZ

Papez integrated brain transection work with ideas derived from comparative studies in the establishment of this anatomic theory of emotion. The final elements constituting Papez's circuit proposal will be examined in the next section.

ANATOMIC MEDIATION OF EMOTIONS (AWARENESS): BASIC COMPONENTS

> "The thesis is that the continuity of the trains of what we call 'thought' which are unceasingly passing through what we call our 'minds' during the periods of what we call 'consciousness' have their neural correlates in a continuity of neural impulses to and fro from thalami to cortex and from cortex to thalami along the multitudinous thalamo-cortical and cortico-thalamic paths in the thalamic fan or radiation."
>
> GRAFTON ELLIOT SMITH, p. 122 (21)

Papez's theory of emotional consciousness embraced two related ideas.

- A central principle of Papez's theory of emotion was that consciousness/awareness arose from reciprocating midline reticulo-thalamocortical circuits (a view which had earlier been suggest by such figures as Charles Dana, Walter Canon, Walter Dandy, Elliot Smith, and Russell Brain—notably the views of Dandy, Dana, and Canon were all cited in his paper). It should also be noted that Papez introduced his paper specifically as a corticosubcortical theory.
- A second assumption adopted by Papez was that emotional experience required a cortical mediator, although emotional expression did not. Papez's put forth this belief on the second page of his paper.

> "The term emotion as commonly used implies two conditions: a way of acting and a way of feeling. The former is designated as emotional expression: the latter, as emotional experience or subjective feeling. The experiments of Bard have demonstrated that emotional expression depends on the integrative action of the hypothalamus rather than on the dorsal thalamus or cortex, since it may occur when the cerebral hemispheres and the dorsal thalamus are totally removed. For subjective emotional experience, however, the participation of the cortex is essential. Emotional expression and emotional experience may in the human subject be dissociated phenomena. Hence, emotion as a subjective state must be referred to the higher psychic level."
>
> 1) p. 726.

The belief that the cortex was necessary for emotional awareness was not Papez's alone. Terms such as sham rage, and pseudaffective reflexes (terms employed by Cannon and Sherrington, see above) indicated that these authorities also felt that higher centers, (i.e., the cortex) had to be involved in enabling the true, as opposed to sham or pseudoemotional experience.

At the time Papez put forth his circuit, neuroanatomic wisdom held that afferentiation of the cortex had to be achieved via a thalamic relay, (the only exception known was the olfactory system). Accordingly, Papez's cortical mediator of emotion—he chose the cingulate gyrus, see below—would also have it's own thalamic relay.

The Cingulate: Cortical Mediator of Emotional Experience

The cingulate gyrus offered several anatomic features that made it well suited to function as Papez's cortical receptive area in emotional consciousness. These features are listed below.

(a) *The Medial Position of the Cingulate*: The medially arrayed cingulate gyrus enjoyed extensive connections, via the cingulate bundle, also mammillothalamic tract, with other *medially* disposed neural loci (parahippocampal gyrus, thalamus also the hypothalamus). Influenced by comparative neuroanatomic theory (see above), Papez accepted a visceral, emotional role for cingulate function.

(b) *Clinical Findings*: Papez cited numerous case reports in which lesions directly involving or impinging on the cingulate-hippocampal continuum caused a significant neurobehavioral disturbance. Examples given included rabies (rabies from the Latin word for fear), in which Negri bodies demonstrate a predilection for the hippocampal formation, also hypothalamus and cerebellum. He also referred to case reports of tumors of the corpus callosum—the behavioral changes were felt attributable to impingement of the tumor on the cingulate. Additionally, Papez pointed to cases of tumors in the vicinity of the anterior thalamic nuclei, also the IIIrd ventricle correlated with behavioral (mental) alterations, as well as disturbances of the level of consciousness.

(c) *Cytoarchitectonics and Connections:* Well aware of the cytoarchitectural work of Cajal, Campbell, also von Economo and Koskinas (1, p. 734) which established the cingulate cortex as more laminated (elaborated) than allocortical districts, and aware that the cingulate was undoubtedly connected to neighboring isocortical districts;[3] also aware that the cingulate was in receipt of thalamic input, Papez was confident the cingulate must be enabling a more complex mechanism.

(d) *Thalamic Relay to the Cingulate*: As noted in the previous section, the 'emotionally receptive' cortex would have to be accompanied by a thalamic relay. The cingulate gyrus had just such a connection—a two step neural connection to the thalamus/hypothalamus. This route included (as the first step) the well-known mammillothalamic tract, also known as the fasciculus of Vicq d'Azyr (described by Felix Vicq d'Azyr back in 1786). Papez drew attention to this two step mechanism, which included, as the second step, the *then* recently discovered anterior thalamic nucleus to cingulate linkage.[4] (22).

> "The cortex of the cingular gyrus may be looked on as the receptive region for the experiencing of emotion as the result of impulses coming from the hypothalamic region, in the same way as the are striata is considered the receptive cortex for photic excitation coming from the retina. Radiation of the emotive process from the gyrus cinguli to other regions in the cerebral cortex would add emotional coloring to psychic processes occurring elsewhere."
>
> 1) PAPEZ, p. 728, 1937

[3] In Papez's day this was a bold concept, since anatomists had not clearly established such connections. MacLean commented that one of Papez's particular strengths was his knowledge that tracts enabling such connectivity (e.g. Muratoff's bundle) might exist (25, p. 266).

[4] Papez referred to cingulate projections from the anterior and anteriodorsal nuclei—the anterior group of thalamic nuclei, not to be confused the ventral anterior thalamic nucleus—is made up of the anterior dorsal, anterior ventral and anterior medial nuclei. MacLean comments that until 1954, probably few anatomists would have suspected the presence of thalamic projections to any part of the limbic lobe *other* than the cingulate. However, the situation changed significantly in the 1950's and 60's with reports of thalamic projections to other limbic cortical sites. MacLean has argued that the association of the anterior group (also dorsomedial thalamic nuclei) with the mesocortex of the cingulate gyrus, remains specific to higher mammals with no clear counterpart in the nonmammalian brain, see chapter 3 for further discussion of this issue.

PAPEZ'S CIRCUIT

Following Papez, we will examine his circuit, by following the route of neural afflux ('incitations' as he put it) into his circuit. Two principle points of neural input were described.

- From various part of the cerebral cortex, via such tracts as the angular bundle, and the subcallosal fasciculus of Muratoff, (relatively obscure tracts to be sure; however, Papez had to rely on the more limited knowledge of connections available in the 1930's) information entered his circuit and then passed to the cingulate gyrus, and hippocampal formation. The hippocampal formation in turn passed the information to the mammillary bodies. Via the mammillothalamic tract neural traffic was conveyed to the anterior thalamic nuclei, and then via the thalamocortical radiation to the cingulate. Flow from the hypothalamus to the cingulate represented to Papez "the stream of feeling." (1, p. 729). Notably, Papez added that the intermingling of cingulate outflow to neighboring cortical districts would add emotional coloring to the psychic processes.
- Neural traffic would also gain entrance to his circuit through the hypothalamus "*It is not yet generally recognized that there are primitive sensory centers in the ventral thalamus, the chief connections of which are appear to pass to the hypothalamus . . . These primitive centers receive certain terminals from various afferent systems.*" (1, Papez, p. 729) Importantly, Papez held forth that hypothalamic input contained a wide variety of both visceral and somatic sensory impressions from peripheral, also internal neural sources. The actual access points to the hypothalamus included such areas as the mammillary peduncle (Papez's expands on this issue, and the reader is referred to his article regarding the particular anatomic linkages he suggested). The neural pathways obtaining these areas (and thus the hypothalamus proper) included, the medial lemniscus, the lateral spinothalalmic tract, also the medial forebrain bundle (this latter white matter pathway is now known to be the brainstems main artery for monoamine traffic). Papez also mentioned (*without elaboration*) possible visceral input from the amygdala to the hypothalamus via the stria terminalis (the function of the amygdala was essentially unknown in 1937). As noted above, from the hypothalamus, impulses would pass, via the thalamus to the cingulate.

SUMMARY

It is important to note that all the elements Papez's circuit had been previously characterized (23,24) although some of the components had only been recently demonstrated, i.e., the thalamacocingulate radiation was described in 1933 (22). Papez's contribution was to put these anatomic elements together and propose a mechanism, embodying features essential (according to his understanding) to an emotional/awareness mediator. These features are reviewed below.

- The circuit would incorporate, in an economical and appealing manner, two vital elements in the elaboration of emotional awareness: a cortical mediator or emotion, and a hypothalamic/thalamic relay to that cortical mediator. In this regard, Papez considered his circuit (indeed he introduced his circuit) as a corticothalamic mechanism of emotion (1, p. 3).
- The circuit would enable endless iterations or reciprocal interactions between the diencephalon and telencephalic cortex, which Papez's also felt to be essential element in the elaboration of emotional consciousness.
- The circuit would also cohere all the medial wall elements including the hypothalamus and seat these elements in a proper (as argued by theorists such as Herrick, & Papez) evolutionary biologic context.

TOWARDS A LIMBIC SYSTEM

Although Papez's circuit is generally (and correctly) described as an essentially closed off and self contained entity, it seems clear, that as actually conceived of by Papez, his circuit was capable of admitting and processing a wide compass of neural inputs, and interfacing with the surrounding midbrain and forebrain.

Formidable as his efforts were to find neural pathways into his circuit, Papez's attempt to expand his circuit beyond the limbic fissure rostrally and hypothalamus caudally could not really succeed. As underscored by MacLean, anatomical connections of core limbic structures with other loci were simply not known in the 1930's, and the methodologies to find these connection not available. The opening up of the borders of Papez's circuit and the concept of a more dispersed limbic consortium would be pursued, employing more advanced techniques, by such figures as MacLean, and Walle Nauta, in the United States, and B.G. Cragg. in London.

It was MacLean who would provide anatomical data in support of an expanded Papez (limbic) circuit in the mediation of emotional behavior. MacLean's efforts, examined in chapter 3, would ultimately lead to the concept of a limbic system. Nauta's contribution is examined in chapter 4.

REFERENCES

1) Papez, J. W. (1937). A proposed mechanism of emotion, Arch. Neurol. Psychiatry, 38, 725–743.

2) Yakovlev, P. (1978). Recollections of James Papez, and comments on the evolution of the limbic system concept, in: Limbic Mechanisms, (K. E. Livingston, and Oleh Hornykiewicz, eds.), Plenum Press, New York, pp. 351–354.
3) Edelman, G. (1992). Bright Air, Brilliant Fire, Basic Books.
4) Sacks, O. (1993). Making up the mind (review of Bright Air Brilliant Fire: On the Matter of Mind), The New York Review of Books, pp. 42–49.
5) Schiller, F. (1992). Paul Broca: Explorer of the Brain, Oxford Univ. Press.
6) Kepler, J. (1975). The Harmonics of the World, trans. C. G. Wallis, Chicago, University of Chicago Press.
7) Broca, P. (1878). Anatomie comparee des circonvolutions cerebrales. Le grand lobe limbique et la scissure limbique dans la serie des mammiferes, Rev. Anthropol. 1, Ser. 2, 385–498.
8) Herrick, C. J. (1933). The functions of the olfactory parts of the cerebral cortex, in *Proceedings of the National Academy of Sciences*, v. 19, 7–14.
9) Zilboorg, G. (1941). A History of Medical Psychology, W. W. Norton.
10) Sulloway, F. (1978). Freud Biologist of the Mind, Basic Books.
11) Kertesz, A. (1983). Localization of lesions in Wernicke's aphasia, in Localization in Clinical Neuropsychology (A. Kertesz, ed.), Academic Press.
12) Finger, S. (1994). Origins of Neuroscience, Oxford.
13) Thatcher, R. W., and John, E. R. (1977). Functional Neuroscience: Volume 1, Foundations of Cognitive Processes, LEA, pp. 117–134.
14) Heath, R. G. (1996). Exploring the Mind Brain Relationship, Moran Printing, Baton Rouge, Louisiana.
15) Woodworth, R. S., and Sherrington, C. S. (1904). A pseudaffective reflex an its spinal path. Journal of Physiology, 31, 234–243.
16) Cannon, W. B., and Britton, S. W. (1925). Pseudaffective medulliadrenal secretion. American Journal of Physiology, 72, 283–294.
17) Shepherd, G. M. (1988) Neurobiology, 2nd edition, Oxford.
18) Romer, A. S. (1962). The Vertebrate Body, 3rd Edition, W. B. Saunders.
19) Herrick, C. J. (1933). Morphogenesis of the brain, *Journal of Morphology*, 54, 233–258.
20) Herrick, C. J. (1948). The Brain of the Tiger Salamander, Ambystoma tigrinum. Chicago: University of Chicago Press.
21) Champion, G., and Smith, G. E. (1934). The Neural Basis of Thought, Kegan Paul, London.
22) Clark, W. E., and Bogan, R. H. (1933). On the connections of the anterior nucleus of the thalamus, J. Anat. 67, 215–226.
23) Meyer, A. (1971). Historical Aspects of Cerebral Anatomy, Oxford University Press.
24) Pribram, K. H., and Kruger, L. (1954). Functions of the "Olfactory Brain", in Annals of the New York Academy of Science, 1954, V. 58: (section C) 109–38.
25) MacLean, P. (1990). The Triune Brain in Evolution. Plenum.

CHAPTER THREE

MACLEAN'S LIMBIC SYSTEM

MacLean was the first to recognise in his 1949 paper "Psychosomatic Disease and the 'Visceral' Brain," that Papez's tentative "proposition" that emotions have a physiologically explorable and anatomically definable machinery was a true discovery and not a "delusion" as had hitherto been widely held. Since then MacLean had demonstrated experimentally, in depth, the epistemic validity of that discovery."

1, p. 351

MacLean's work represents an island of accessible and wide-ranging generalizations in an ocean of abstruse and arcane technicalities. Drawing on themes that have been influential in our culture for generations, it provides an ambitions synthesis of biology and psychology that reinforces everyday or commonsense perceptions of human beastliness and offers apparently authoritative judgments concerning human health and happiness.

(2)

INTRODUCTION TO MACLEAN'S CONTRIBUTION

Paul MacLean figured preeminently in both the popularization, and the further development of Papez's work on the anatomy of emotion. In order to gain an appreciation for MacLean's efforts in elucidating the anatomy and function of the limbic brain, we have divided his rather extensive body of work into three areas. The first section examines MacLean's role as the disseminator and popularizer of Papez's proposal in the middle decades of this century—importantly, this was a role played out during the height of the Freudian dispersion, and the heyday of black box behaviorism, an era in which psychiatric medicine was lukewarm (if not frankly hostile) to anatomically determined theories of behavior. The second section (entitled, from Papez's circuit to MacLean's limbic system) will overview MacLean's endeavor to establish periallo- and neocortical links to Papez's circuit (i.e., to expand the confines of the limbic apparatus beyond the confines of the limbic fissure). It is this endeavor that would result in his limbic system proposal. The last sections of the chapter examine MacLean's efforts to place the mammalian limbic brain in an evolutionary, and comparative neuroanatomic context.

The Dissemination of Papez's Proposed Mechanism of Emotion

It is helpful to place MacLean's proposal of a *limbic system* (3,4) in the context of Papez's proposed limbic circuit (5). Fortunately this undertaking is facilitated by MacLean's own comments regarding his early endeavor in the area of the anatomy of emotion, and the intellectual heritage he shared with Papez.

> "My introduction to him [Papez] happened in this way: In 1947, I had received a USPHS Fellowship for working with Dr. Stanley Cobb . . . Through Dr. Cobb I arranged to conduct some electroencephalographic research in the Brain Wave Laboratory . . . There I developed an improved electrode for recording the electrical activity at the base of the brain. This work led to a diagnostic electroencephalographic examination of patients with so-called psychomotor epilepsy patients with this form of epilepsy experience one or more of a wide range of vivid emotions at the beginning of their seizures, i.e., during the aura. In combined recordings of the basal and standard electroencephalogram, Arellano and I found that in some cases the maximal bioelectrical disturbance (manifest by "spike" potentials) occurred in the medial basal region of the temporal lobe, suggesting an origination in or near the hippocampal formation.
>
> About the same time I happened upon Papez's neglected paper on emotion and was struck by its obvious relevance to the emotional symptomatology of our patients in our study. In conjunction with their emotional feelings, patients with psychomotor epilepsy may experience symptoms identified with one of more of the sensory systems. Except for olfactory connections, there was almost no indication in Papez's paper how the hippocampal formation would receive information form other sensory systems. Papez had noted that an abstrusive structure known as the subcallosal bundle (Muratoff's bundle) might provide an associational

> link between the general cortex and the hippocampal formation. The general cortical areas were located in the frontal, cingular, and parietal regions. With only these inputs, how was one to explain the various visceral, gustatory, somatic, auditory, and visual symptomatology that patients with the above kind of epilepsy experienced at the beginning of their attacks? It was this kind of question that took me to Ithaca in 1948 to discuss the problem with Dr. Papez."
>
> (6, pp. 265–266)

This citation reveals two related insights into MacLean's early effort: he was a firm believer in the concept of a broadly conceived central mediator of emotional behavior, and he felt confident (based on clinical experience, EEG findings, and newer findings emerging from tract tracing experiments) that the central mediator was far more extended than Papez's original description suggested. Consequently, MacLean was determined extend the compass of anatomic based explanation beyond Papez's circuit. MacLean continues his commentary and reveals the direction he would help establish for future developments in limbic brain research.

> "On the strength of my visit to Papez and accumulated new information since his original article, I prepared a paper for presentation at a departmental seminar that in 1949 was published under the title, "*Psychosomatic Disease and the 'Visceral Brain.' Recent Developments Bearing on the Papez Theory of Emotion.*" Figure 3 of that paper shows the sketch I used to summarize suggestive evidence of overlapping inputs to the hippocampal formation from all of intero- and exteroceptive systems."
>
> (6, p. 266)

The 'visceral brain' paper (referred to above) is of particular significance for at least two reasons: to a large segment of the behavior neuroscience community at mid century this publication helped to both endorse, and popularize Papez's decade old proposal—it is this paper that Yakovlev is referring to in his epigraphs to both this and the last chapter. Furthermore, it is in this paper that we first gleam the germ of MacLean's limbic system concept a concept which at its most simplest posited a consortium of dispersed yet interactive neural components extending beyond, but intimately connected with Papez's circuit (the components of Papez's circuit restricted mainly to the allo- and mesocortex). Forty years after its publication, MacLean made the following comment on the merit of this important publication.

> "The visceral brain paper was perhaps significant for introducing a few new ideas. First of all it suggested how a phylogenetically old part of the brain, found as a common denominator in mammals, might receive information from all the sensory systems. In regard to the hippocampus itself, this would indicate that it was not an autonomous little factory of its own, manufacturing the raw materials of emotion out of thin air."
>
> (6, p. 266)

THE VISCERAL BRAIN VS THE LIMBIC SYSTEM

MacLean explains why he initially chose the term visceral brain to refer to the limbic area. *"I used the expression* ***visceral brain*** *as a means of avoiding the narrow implications of the term* ***rhinencephalon****. In its original 16th century meaning,* ***visceral*** *applies to strong inward feelings and implicitly the accompanying visceral manifestations."* (6, pp. 266–267, emphasis MacLean's). MacLean, it seems, advanced the term in the hope that it might help point the way to anatomic correlation of emotional experience: in this regard it stood in contrast to such eponymous descriptors as Papez's circuit, and Broca's great limbic lobe, see Table 1.

However, the possible advantages of the term visceral brain were not realized; MacLean would abandon it within three years. The term was criticized for two reasons. Karl Pribram, an early and close collaborator of MacLean's cast doubt on the existence of a distinct visceral brain. Pribram argued that areas of the hemisphere generally considered somatic as well as those classically considered to be limbic/visceral both contributed to visceral function; it was also evident, Pribram argued, that areas traditionally 'assigned' to the limbic lobe contributed to somatic function—to Pribram there simply was not sufficient data to support the presence of a discrete visceral, as opposed to a somatic brain. Pribram's comment, also MacLean's response, were appended as a discussion to the ORIGINAL article containing MacLean's proposal of the term, limbic system (4). MacLean accepted Pribram's argument against any absolute distinction between a visceral as opposed to a somatic brain, yet held to a localizationistic position and argued that sufficient data did exist to support the presence of a part of the brain that functioned, as he put it, *"with an 'eye' more to interpreting and giving expression to the internal (visceral) needs of the body, than . . .'non-felt' ideational functions."* (4, p. 417)

A second problem with MacLean's use of the term visceral was that it came into conflict with a more restricted usage common amongst anatomists, also physiologists. MacLean commented on this conflict, *"I found that the term created misunderstanding because in physiological parlance the word visceral applies only to glands and hollow organs, including the blood vessels."* (6, p. 267). MacLean felt that his visceral brain was certainly engaged in more than just the adjustments of smooth muscle lined lumina (the restricted anatomic usage). MacLean would later note that it was the conflict with basic anatomic usage, and *not* the more complex issue of localizability versus non localizability of emotional behavior that lead him to abandon it and adopt Broca's 75 year old terminology. MacLean comments on the resurrection of Broca's term.

> *"I resorted to Broca's descriptive term* ***limbic*** *and used the expression* ***limbic system*** *when referring to the cortex of the limbic lobe and the structures of the brainstem with which it has primary connections. This explains how the term limbic system was introduced into the literature in 1952."*
>
> *(6, p. 267 emphasis MacLean's)*

From Papez's Circuit to MacLean's Limbic System

MacLean understood that Papez's circuit could explain in anatomical terms, some, but far from all of the clinical (experiential) material in the setting of complex seizures. The road map Papez provided simply lacked the anatomical data—the actual links between associational (neocortex), and Papez's core circuit elements—that MacLean assumed must be present to explain the broad compass of emotional experience, and the associated electroencephalographic (EEG) findings encountered in the neuropsychiatric setting. In the following sections, we will follow MacLean's use of three strategies: strychnine neuronography (a neuroanatomical research technique widely used at mid century), electroencephalography, and correlative clinical phenomenology in his pursuit of these connections.

Contributions from Neuronography to the Development of the Limbic System Concept

Results derived from strychnine neuronography conducted in the 1940's and in the early 1950's, contributed significantly to MacLean's confidence that he was in possession of firm anatomic data that Papez's lacked—evidence for neocortical connection to Papez's circuit. Based on such evidence, in concert with EEG findings, also clinical studies, MacLean would put forth his limbic system proposal.

Strychnine studies conducted by MacLean, displayed in figure 3.1, suggested to him that the posterior orbital, anterior insular, temporal polar, and piriform areas were all extensively and reciprocally connected—and therefore seemingly distinguished from neighboring cortical regions. MacLean combined the result of this study with earlier neuronographic reports to establish the presence of connections from the caudal orbitofrontal area to the posterior hypothalamus. MacLean interpreted the results of these two studies to mean that the frontotemporal cortical area stood in rather close anatomic, and physiologic contact with the limbic core. MacLean championed this newly found anatomic evidence of neocortical to allocortical to hypothalamic connections in a rather bold fashion on the first page of his limbic system paper[1].

> "... we now know that the posterior orbital, anterior insular, temporal polar, and pyriform areas are all reciprocally connected ... it is also known from the strychnine studies that the entire frontotemporal region fires into the amygdala and the rostral hippocampal, which in turn have been shown by classical anatomical methods to have a heavy projection on the hypothalamus, septal region and parts of the basal ganglia."
>
> (4, p. 407: emphasis added)

With the publication of MacLean's limbic system paper, the constraints on limbic circuitry encumbering Papez's were broached.

The Contribution of Clinical and EEG Findings to MacLean's Limbic System Concept

> "We cannot go right on with the psychology, not with the anatomy, nor with the pathology of our subject. We must consider now one and now the other, endeavoring to trace a correspondence betwixt them."
>
> HUGHLINGS JACKSON (8)

In addition to newer anatomic and physiologic findings strongly suggesting frontotemporal linkage to Papez's circuit, MacLean (as Papez

[1] Note MacLean's inclusion of the amygdala in his 1952 (limbic system) publication from which this citation is taken. His decision was influenced by the Kluver-Bucy experiment, also chemical and electrical stimulation studies linking the amygdala anatomically and functionally to Papez's core. Earlier mention of the amygdala as a subcortical nucleus in the limbic organization appears to date to Cajal, (4, p. 407fn). As noted in chapter 2, Papez mentioned the amygdala nucleus without elaboration in his paper.

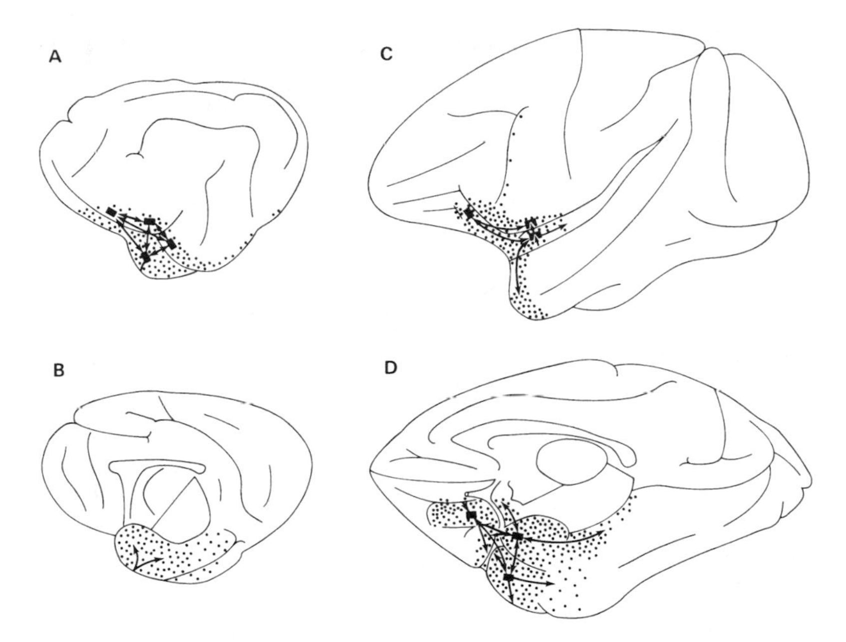

Figure 3.1. This figure summarizes the results of neuronographic studies carried out by MacLean and Pribram in the cat (although now only of historic interest, when first developed by Dusser de Barenne, and McCulloch, neuronography provided an early physiologic means to investigate the arrangement of associational and commissural systems). These studies helped to identify connections of core limbic structures with orbitofrontal stations. In this technique a small rectangle of paper soaked in 1% strychnine was deposited at selected sites (rectangular areas); subsequent mapping of propagated spike potentials revealed interconnected cortical areas. In this figure, the stippled areas indicated reciprocal linkage. Examination of the stippled areas suggested an abundance of connections between the caudal orbitofrontal cortex and the rostral part of the temporal lobe (including superior, middle, and inferior temporal gyri). Most of these connections were routed via the uncinate fasciculus. MacLean combined the results of his studies, with earlier evidence (also neuronographic): together the studies demonstrated caudal orbitofrontal cortical connection with the hypothalamus thus establishing that the orbitofrontal area partakes in the limbic system. Within ten years, fiber degeneration studies employing the Nauta-Gygax staining method (a method, which provided a far higher degree of reliability over strychnine studies) affirmed MacLean's main contention regarding orbitofrontal to limbic connections (7). Figure reprinted with permission from (6, p. 294).

before him) would look to clinical correlation in support of his anatomic theories. The following excerpt from MacLean's limbic system paper provides an appreciation of his use of behavioral (often subjective, experiential) correlates.[2]

> "From the study of epileptogenic foci in the frontotemporal region, one gets a clue regarding the subjective functions of this part of the brain . . . At this point I shall introduce the reports of three patients I have seen recently to emphasize the epigastric aura and its associated sensations represent raw, poorly differentiated, and impersonalized feelings. One of these patients relates to his epigastrium a feeling of sadness and of wanting to cry. This is accompanied by a welling up of tears and sensation of hunger. A second patient refers to his epigastrium a feeling of wanting somebody near him. And finally I refer to a young man who senses in his stomach a feeling of fright that carries with it the conviction that someone is standing behind him. If he turns his head to see who it is, the feeling of fear becomes intensified. The EEG findings in these patients point to a epileptogenic focus somewhere in the frontotemporal region."
>
> The thing I wish to stress in all these examples is the lack of identification of the feeling with any specific event or person. In short, it is "feeling" out of context. The sadness felt in the epigastrium pertains to no particular situation. The feeling of need of company is not related to particular persons . . . Here then is raw, poorly differentiated, and impersonalized feeling, viscerally related, and presumably stemming from discharges in the frontotemporal region."
>
> (4, pp. 412–413, 1952, emphasis added)

Anatomic Correlation in the Development of the Limbic System Concept—A Different Clinical Imperative

The abiding belief that clinical findings could be correlated with neocortical connections to the core components of Papez's circuit was not

[2] The search for neuroanatomic correlates of visceral experience has a long tradition dating to the 19th century. For a review of the early history of this research see (9, pp. 5–20). For a short overview of efforts to relate visceral experience to brain location in the middle decades of this century, see (10, pp. 11–14).

restricted to MacLean, also Papez and their search for anatomic correlates of emotional behavior. In the decade following MacLean's proposal of a limbic system, B. J. Cragg at University College London pointed to the lack of known anatomic connections to explain the rather robust and dramatic loss of memory occurring in the setting of hippocampal injury. Cragg, who was clearly referring to the case of H. M. wrote, "*In view of the apparent importance of the hippocampus in the formation of new memories . . . it is surprising that the neocortical connexions of the allocortex are not more prominent.*" (11, p. 352).

The case of H. M. is considered a landmark in the development of our knowledge regarding memory function. In 1954, in a paper entitled, The Limbic Lobe in Man, (12) the neurosurgeon William Beecher Scoville, reported on the then unexpected finding of severe and persistent loss of memory after bilateral removal of the medial parts of the temporal lobes—this paper contained the first mention of this famous case (for an extended narrative, and scientific review of this case see Hilts (13). To the experimental neuroanatomists, as well as to Scoville, the case of H. M. presented a puzzle. A structure of such evident importance to the mediation of memory as the hippocampal formation (as revealed in the tragic case of H. M.) would presumably be expected to have strong anatomic connections with the sensory systems – the sensory systems which provided the material to be remembered; however at the time Scoville performed the surgery, such connections were not characterized.

Employing the Nauta-Gygax technique, Cragg helped establish the presence of associational cortical to limbic core pathways: specifically he demonstrated suprasylvian parietal cortex links to the cingulate; also projections from the prefrontal cortex to the temporal cortex (with subsequent passage to the entorhinal area). Presumably such connections are instrumental in information delivery to and ultimately memory consolidation by the hippocampus. H. M.'s inability to remember new information received (in part) an anatomic explanation.

On the surface, Cragg's and MacLean's search for limbic connections appeared motivated by entirely disparate clinical concerns (memory function in contrast to emotionality); however, it should be noted that the role of the limbic brain in memory process and in affective behavior undoubtedly have a close underlying interrelationship.

> *"The hippocampus represents a major limbic structure that has been shown to be functionally involved in the initiation of locomotor movements in exploration and in the processing of spatial memory. Excitatory glutamatergic hippocampal outputs project to the accumbens where dopaminergic inputs from*

the VTA also converge. Conceivably, hippocampal signals coding for spatial orientation, and VTA signals coding for reward significance, can be integrated into motor behavioral output."

(14, p. 194)

Summary

As early as 1949, MacLean felt that the presence of convergent inputs (intero- and exteroceptive) to the allocortical limbic core had been sufficiently established to support the contention that there was a consortium of neural structures, closely interfaced to, but also extending beyond Papez's circuit—a neural macrostructure which constituted a limbic system, the term he coined (4). MacLean's limbic system brought with it the promise of an explanatory power in the clinical setting of epilepsy and psychiatric illness unavailable previously.

Table 1 summarizes the main components of the limbic architecture as put forth by Broca, Papez, and MacLean. The inexorable growth of the limbic brain consortium would, depending on ones perspective, either strengthen or weaken the concept of the limbic brain.

AN AMBITIOUS SYNTHESIS: THE LIMBIC BRAIN IN AN EVOLUTIONARY BIOLOGIC CONTEXT

After the completion of his early papers in which he championed and extended Papez's proposal, MacLean engaged in a enterprising endeavor to place the limbic system of mammals in a broadly conceived evolutionary context; an endeavor that might shed light on the observational findings dating to the early 19th century of a constant (and in higher mammals a rather robust) centrally disposed circumannular convolution. As noted by Durant, this endeavor was an ambitious effort indeed, and not easy to summarize. Fortunately, for the purposes of this chapter, we can turn to MacLean for a summary statement of his principal findings.

"It is beginning to appear on the basis of comparative neurobehavioral studies that the cingulate subdivision of the limbic system is implicated in three forms of behavior that characterize the evolutionary transition from reptiles to mammals—namely (a) nursing, in conjunction with maternal care; (b) audiovocal communication for maintaining maternal-

Table 3.1. Major Components of Limbic Architecture: Broca (1878) through MacLean (1952)

Eponym	Cortical Structures	Subcortical Structures
Broca's Limbic Lobe (1878)	Olfactory lobe (olfactory bulb, tracts). Gyri: Subcallosal, fornicate (cingulate, parahippocampal), hippocampus, and Ammon's horn (see chapter 1)	Paraolfactory, paraseptal areas. The diagonal band, also the *carrefour*. The *carrefour*, a long abandoned term, was an area of multiple communications located at the origin of the callosal gyrus: the *carrefour* is roughly equivalent to the paraolfactory area; other terms for this area include the paraterminal body).
Papez's Circuit (1937)	Similar to Broca Plus: Obscure tracts such as Muratoff's bundle (see chapter 2).	Similar to Broca: *Plus*: The hypothalamus*. The anterior nucleus of thalamus, and the amygdala. The amygdala suggested to carry visceral information to the hypothalamus (5, p. 742).
MacLean's Limbic System (1952)	Similar to Papez *Plus*: Frontotemporal associational cortex; and perihippocampal centers arrayed around the limbic fissure	Similar to Papez *Plus*: Amygdala

* The term hypothalamus was unknown to Broca: according to Schiller, Edinger introduced the term in the 1890's (15, p. 248).

offspring contact; and (c) playful behavior. Significantly, the cingulate gyrus and its subcortical connections appear to have no recognizable counterpart in the reptilian brain "

(16, p. 23)

To the traditionally accepted canonical features of mammalian biology, such as fur, endothermy, placentation, and the possession of mammary glands, MacLean would add that mammals are also distinguished by the presence of an elaborated limbic system (with circuit elements, such as the thalamocingulate, specific to the mammalian grade[3]). As noted above, MacLean's idea on limbic brain evolution covered a wide multidisciplinary compass; to better appreciate his accomplishment (also the limitations of his effort) would require an introductory disquisition into paleontology, comparative neuroanatomy, comparative behavior, and a discussion of his triune brain theory: a worthy undertaking indeed, but well beyond the space limitation of this chapter. We will therefore settle, in the following subsections, with a highly abbreviated presentation of some basic concepts: helpful as an introduction to MacLean's perspective, but only intimating at its scope.

Phyletic Considerations

> "Before returning to contemporary history, it is necessary to delve into ancient history and review how three forms of behavior that most clearly mark the evolutionary dividing line between reptiles and mammals found representation in the cingulate gyrus."
>
> 16, pp. 1–2

The first terrestrial radiation of backboned animals in the form of the amphibians occurred about 350–400 million years ago. The lack of two evolutionary adaptations, a watertight skin, and the amniote egg prevented the amphibians from a more complete mastery of the land. The acquisition of these two characteristics was a principal achievement of the amniotes (reptiles, birds, mammals).

In possession of the two aforementioned adaptations, the reptiles became the first fully terrestrial vertebrates. The amniote egg could be deposited on land and mature to the adult form without passing through the risky aquatic larval stage. The egg yolk is the only food the reptilian mother provides for its baby. From this food source, the animal must develop to the point of independence, capable of seeking food for itself. Parental care of the hatchlings is generally very limited in reptilian forms.

Among the very earliest reptiles there appeared a subclass that bridged the gap between the ancestral reptiles and the protomammals. Oldest of this subclass were the pelycosaurs. The distant descendants of the pelycosaurs are the mammals, (which includes of course, us).

---

[3] In the language of cladistics, MacLean would claim that the thalmocingulate division is a derived feature seen only in mammals.

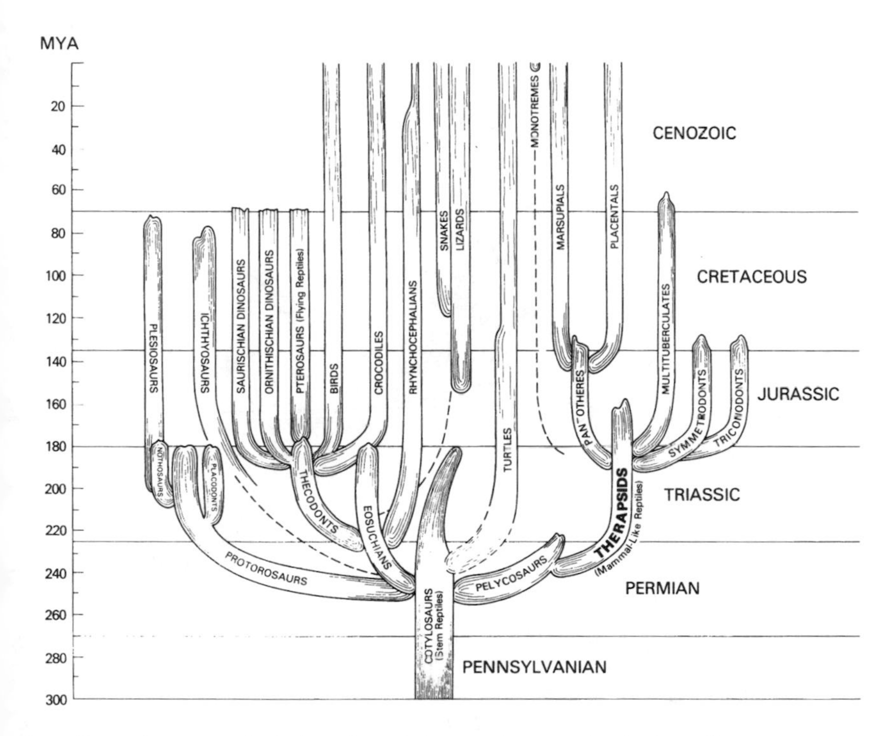

Figure 3.2. Cactus tree depiction of the phylogeny of mammals. MacLean directs our attention to the fact that the pelycosaurs (those sail backed creatures which, are *NOT* dinosaurs at all but the distant ancestors of mammals) have a *DIFFERENT* branch point off the stem reptile than the reptilian forms we see today. The strange sail supported by the pelycosaurs may have represented early attempts at better body temperature regulation—a step perhaps on the way to warm-bloodedness. MacLean underscores the point that it is only in the pelycosaur > therapsid > mammalian line that one sees the full efflorescence of the limbic brain, and its associated triad of family behaviors. Notably, the pelycosaur > protomammal branch occurred in the Permian very soon after the appearance of the stem reptile: this was long *BEFORE* the advent of the dinosaurs. Reprinted with permission, (6, p. 34).

An Introduction to the Triune Brain Concept[4]

Placed in its proper paleontological and phylogenetic context see figures 3.2 & 3.3, the triune brain concept can now be introduced (and ultimately subject to question). Once again we turn to MacLean for a summary of his main contention.

---

[4] MacLean's formulation of the triune brain concept occurred at a time when advances in anatomic tract tracing had largely overthrown the traditional view of a mammalian

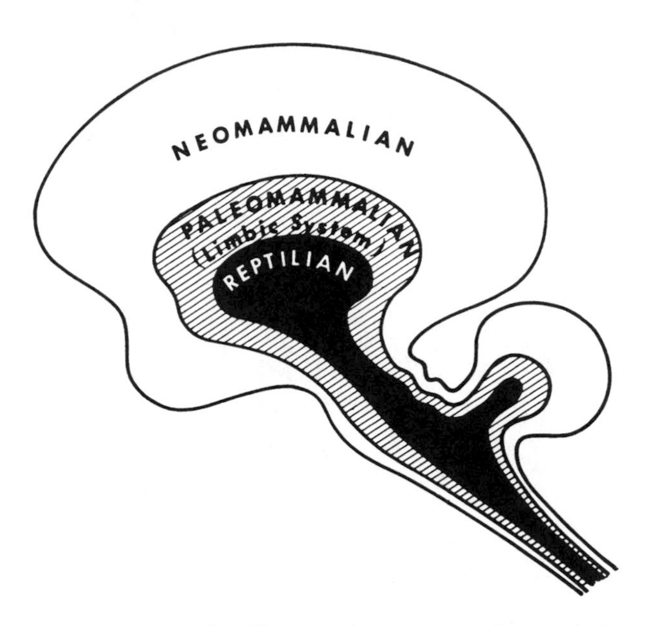

Figure 3.3. MacLean's iconography of his triune brain concept. Over evolutionary time, each division of the human forebrain expanded along the lines of three basic formations that anatomically and biochemically reflect an ancestral relationship respectively to reptiles, early mammals, and late mammals. Reproduced with permission from (6).

> "The comparative study of brain of extant vertebrates, together with an examination of the fossil record, indicates that the human forebrain has expanded to its great size while retaining features of three basic evolutionary formations that respectively reflect ancestral commonalties (sic) with reptiles, early mammals, and late mammals. Markedly different in structure and chemistry, and in an evolutionary sense eons apart, the three major formations constitute, so to speak, an amalgamation of three brains in one—a triune brain. These considerations, together with neurobehavioral findings, attest to marked disparities in the psychological functioning of the three formations. A special complication presents itself in

hemisphere compromised of three largely independent domains: the cerebral cortex, the striatum, and the limbic lobe, see chapter 1. MacLean's triune theory would seem anachronistic even at the time it was put forth. Addressing various challenges to his concept, MacLean would write, "*My emphasis on these distinctions seems to have led some writers to conclude that higher mammals are under the control of three autonomous brains. It was to guard against such an interpretation that . . . I began to use the expression "the triune brain". Triune, a concise term derives letter by letter from the Greek. If these formations "are pictured as intermeshing and functioning together as a triune brain, it makes it evident that they cannot be completely autonomous, but does not deny their capacity for operating somewhat independently.*" (6, p. 9)

> human beings because of evidence that the two evolutionarily older formations do not have the capacity for verbal communication."
>
> (17, p. 3)

The three divisions of MacLean's triune brain are discussed in the subsections below.

The Protoreptilian Formation: The R-Complex At the base or core of the forebrain in reptiles, birds, and mammals is the striatal ganglionic complex, see figure 3.4. These structures are predominantly located in the telencephalon but are extensively linked to adjoining diencephalic satellite gray loci. The deep gray structures include the corpus striatum (caudate and putamen), the olfactostriatum (a somewhat dated term used by Herrick to refer to the accumbens and olfactory tubercle), and lenticular nucleus. Lenticular means lens shaped (the putamen, laterally, and pallidus, medially, assume a lens shape appearance). Underscoring the great antiquity, and the ubiquity of the R-complex, MacLean writes, "*. . . the German neurologist Edinger commented, these ganglia must be of 'enormous significance' for otherwise they would not be found as a constant feature in the vertebrate forebrain.*" The R-complex is notable for its histochemical staining for acetylcholinesterase and dopamine.

The Protomammalian Formation: The Limbic System The forebrain assembly identified with early mammals corresponds to the so-called limbic system. In higher mammalian forms, this includes the rather robust circumannular convolution of Broca; also the associational cortex and the structures of the brainstem with which it has primary connections. MacLean points out that unlike the R-complex which is a fundamental core seen in the brain of all vertebrates, the paleomammalian/limbic brain is rudimentary or only partially represented in reptiles and birds.

Ariens-Kappers, who conducted a wide ranging series of comparative studies in the early 20th century made the following comment on the salience of the limbic brain which clearly echoes Broca's description, and lends observational support to MacLean's view.

> "It is remarkable that the sagittal fissures, which bound this region dorsally, remain constant to a considerable extent throughout the mammalian series, not only morphologically but also in their relation to the underlying cytoarchitectonic fields."
>
> (18, p. 1592)

The Neomammalian Formation This is the neocortex, the outermost and largest of the hemispheric realms, which enables the highest level of sensory analysis and integration, as well as certain aspects of motor

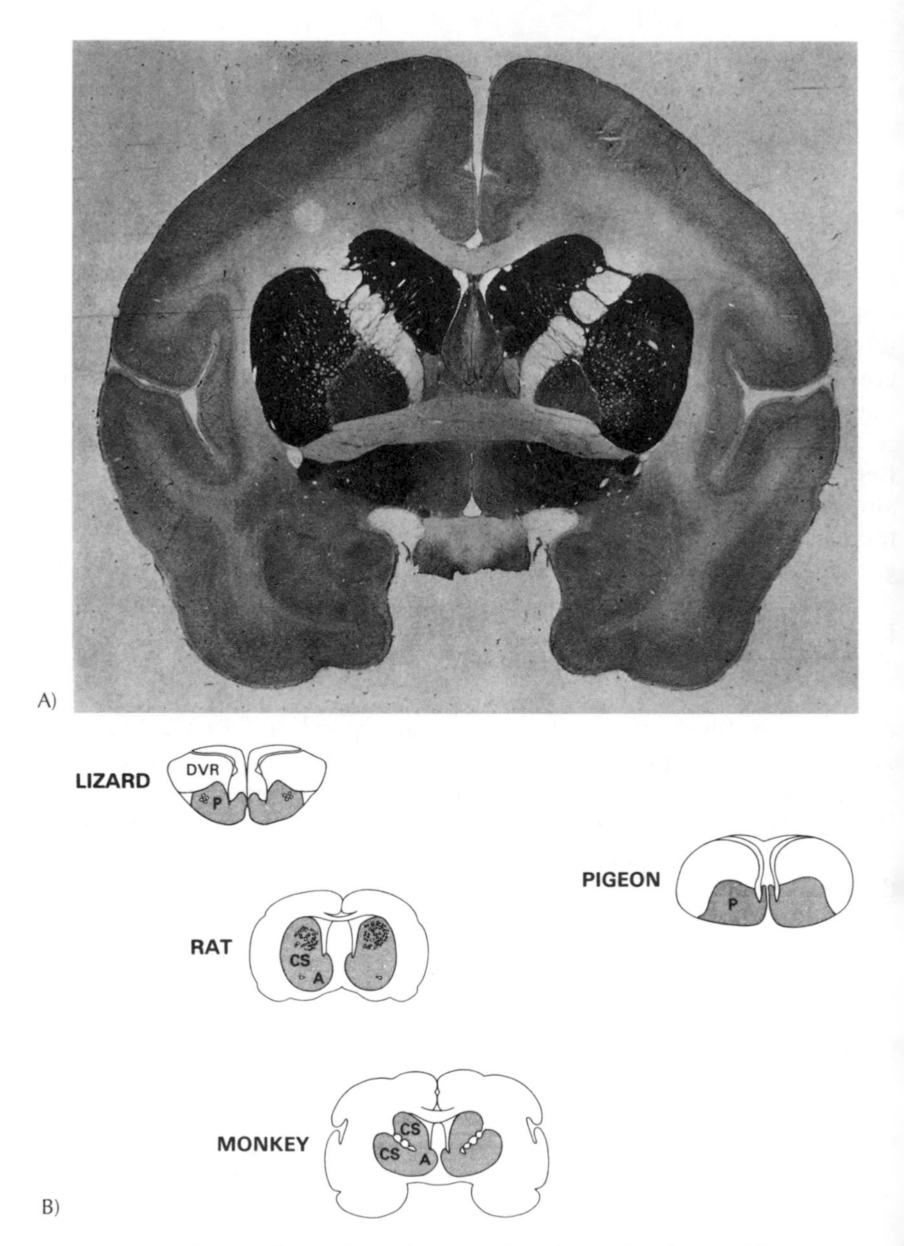

Figure 3.4. A) The striatal complex of the squirrel monkey is sharply set off from the rest o[f] the hemisphere by its rather dense staining for cholinesterase. This is the hemispheric core— MacLean's R-complex. B) The R-complex (MacLean's term for the striatal core when used in [a] comparative context) in a reptile, rodent, primate, and bird: the R-complex is a commo[n] denominator of the brains of all terrestrial vertebrates, reproduced with permission, (6, p. 39)

integration. The neocortex is regarded as a principal substrate of thought processes. In terms of the growth ring hypotheses, the neomammalian formation represents an elaboration or outgrowth from the allocortical and peirallocortical investments.

MacLean's Triune Brain; and Freud's Tripartite Theory of the Mind

As early as 1949 MacLean spoke of parallels between his idea regarding bipartite brain anatomy and function and Freudian theory. MacLean suggested that in humans, the visceral brain stood in a similar relationship with the enormously expanded neocortex as Freud's unconscious mind or "id" stood with relationship to conscious mind or "ego".

> *"Considered in the light of Freudian psychology, the visceral brain would have many of the attributes of the unconscious id. One might argue, however,* ***that the visceral brain is not at all unconscious (possibly not even in certain stages of sleep), but rather eludes the grasp of the intellect because its animalistic an primitive structure makes it impossible to communicate in verbal terms."***
>
> *(3, p. 348, emphasis MacLean's).*

MacLean argued that communication between the visceral brain and the neocortex is hampered by the fact that the visceral brain is prosematic or nonverbal. Prosematic communication includes any kind of nonverbal signal—vocal, bodily or chemical (6, p. 11). In this regard the prosematic visceral brain mirrored the non-verbal primary process of Freud's id. The lack of communication, according to MacLean could result in certain psychopathic states in which emotions built up in the visceral brain to the point where they cascaded through the autonomic nervous system in what he called "a kind of 'organ language' ". (3, p. 350).

Comparative Behavioral Profiles Analysis: Contrasting Reptilian (R-Complex) and Mammalian (Paleomammalian/Limbic) Mediated Behavior

"Origin of man now proved—metaphysics must flourish—He who understands baboon would do more towards metaphysics than Locke"

(19, p. 539)

"We should have a better insight into the "brain of a snake" and "the appetite of a bird"—and we just might reach a better understanding of ourselves."

(20, p. 43)

"Because of this unique family-related triad, one might say that the history of the evolution of the limbic system is the history of the evolution of mammals, while the history of the evolution of mammals is the history of the evolution of the family."

(6, p. 247)

Based on his studies of reptilian behavior (employing a seminatural ethological experimentation), MacLean held that the R-complex mediated stereotyped behaviors related to the establishment of home territories: finding shelter, and food, also sexual and aggressive displays. Importantly, all display and communicative behaviors mediated by the R-complex occurred at the non-verbal, or preverbal level. This nonverbal communication was referred to by MacLean as prosematic communication—prosematic (G. *sema*, refers to mark, token, or sign; when combined with pro it means a rudimentary sign or token).

According to MacLean's analysis one could identify many forms of behavior *common* to reptilian and higher mammalian forms: this meant that in animals as diverse as lizards and primates, the R-complex plays a basic role in non verbal communication, and other primal behaviors. However, of central significance to MacLean's theory regarding the limbic elaboration was his finding that the mammalian triad of family behaviors was for the most part *absent* in the reptilian form. In MacLean's theory, simply put, this triad of behaviors (nursing in conjunction with infant care, audiovocal communication to maintain maternal infant contact, and play behavior) was *outside* the competency of the R-complex: and would be embodied in the paleomammalian (limbic) consortium. In the following sections certain aspects of mammalian biology that seem to accord with MacLean's theory are briefly mentioned.

Placententation, Viviparity, Mammary Glands and Increased Maternal Care

"Most lizards lay their eggs go away and leave them to hatch on their own. The hatchlings come into the world prepared to do everything that they have to do except to procreate. Infant mammals on the contrary are totally dependent on nursing and maternal attention. Any prolonged separation from the mother is calamitous, a situation that again calls

> attention to the importance of the separation call for maintaining maternal-offspring contact."
>
> 16, p. 6

The amniote egg of the reptile bundled a small representation of the Paleozoic seas in a relatively protected environment yet predation would in the course of time exact its own peril on hatchlings of reptiles. The evolutionary variant of placental breeding allowed the mammalian young to remain in the mother's body for a much longer period.

Through such new organs as the placenta and mammary glands, maternal nurturance became extended and very intimate over time. Protracted maternal care required a highly integrated set of neurovisceral, neuroendocrine, and neurobehavioral processes, which MacLean would argue relate most directly to the mammalian limbic elaboration, see figure 3.5.

To support his contention that the limbic system, particularly the anterior cingulate, relates directly to maternal behaviors and play, MacLean pointed to earlier studies correlating cingulate ablation with severe parental deficits done in rats. He also engaged in his own ablation studies to provide further support that integrity of the cingulate limbic cortex relates directly to parental behaviors.

Audiovocal Communication—The Development Of Middle Ear Bones

> "Seldom have we met a case in any part of animal organization, in which the original form of an early [embryological] condition undergoes such extensive change as in the ear bones of mammals. We could scarcely believe it . . . Nevertheless, it happens in fact."
>
> 21, p. 103: C. B. Reichert as cited in S. J. Gould

> ". . . the possibility suggests itself that the so-called *separation cry* (MacLean's emphasis) may be the oldest and most basic mammalian vocalization."
>
> (6, p. 397)

MacLean draws our attention to the fact that reptilian forms engage in little or no audio vocal communication between mother and child, eggs are laid and quite commonly the parent moves on. Even if reptiles 'wanted' to engage in vocal communication between generations it would be impossible—reptiles are probably mute and their hearing is limited. Many mammals, particularly higher primates are quite noisy: and enjoy a complex

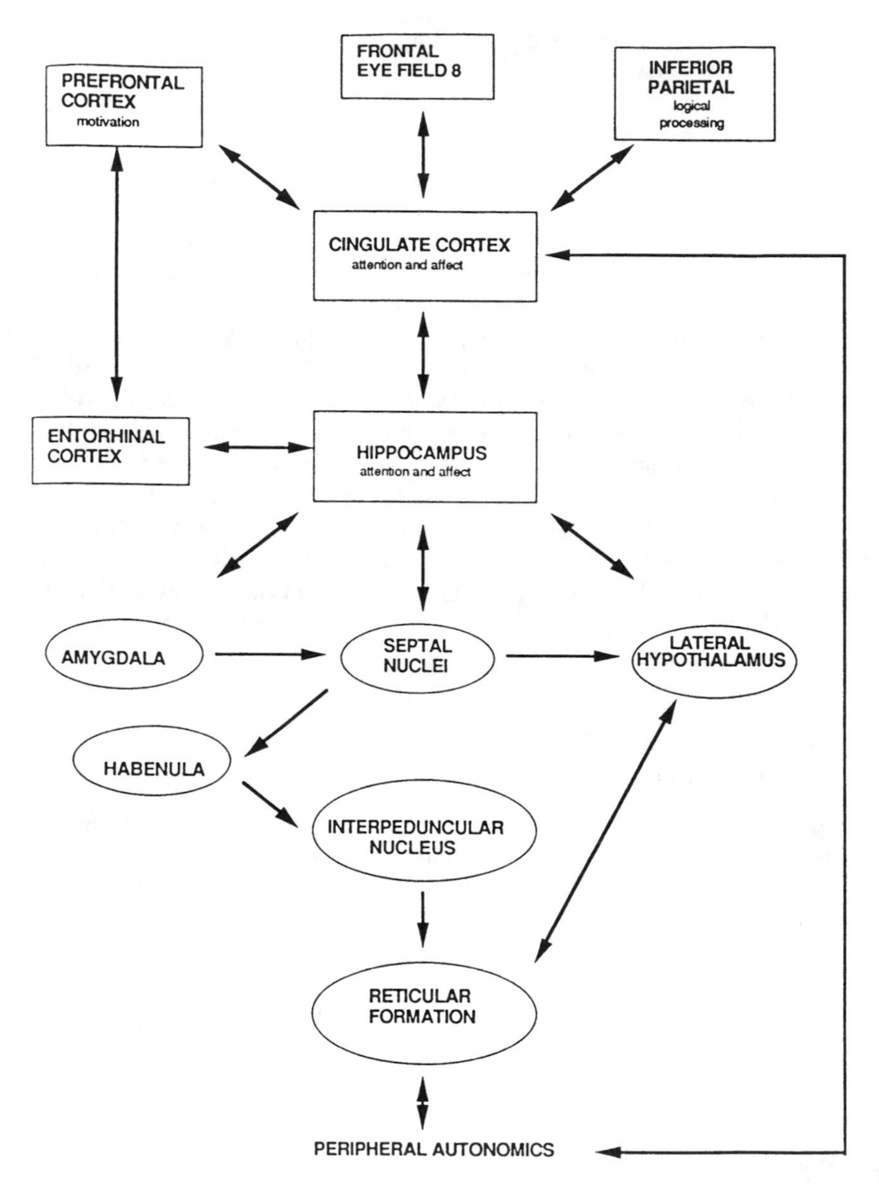

Figure 3.5. This neuroanatomic schematic on the development of the corticolimbic system illustrates the evolutionary acquisition of many of the main limbic structures and their interrelationships. The figure underscores how the paleomammalian limbic structures are overlaid and interfaced with the earlier vertebrate neural (including core limbic) apparatus. Subcortical areas shown in ellipses are found in early vertebrates; mammalian aspects of the limbic system (corticolimbic circuitry) are indicated in the rectangles. Reproduced with permission from (22).

of three middle ear bones (evolutionary processes transformed gill bar elements into jaw bones, and then transformed the jaw bones into the middle ear bones) to convey the sound. Occurring parallel to the osteologic changes elaborating the mammalian middle ear (as first characterized by Reichert) were changes in the neurologic and neurobehavioral anatomy referenced to the limbic system (as suggested by MacLean) enabling the maternal infant audiovocal communication.

Regarding the possible coevolution of the limbic mechanism, and the middle ear organ, MacLean writes,

> "Interestingly, the elaboration of this so-called 'limbic lobe' coincided with the appearance of audition, vocalization, maternal nurturance, and separation calls by the young, suggesting that the integration of visceral responses with cortically mediated behaviors might have facilitated the protection of young offspring."
>
> 22, P. MacLean, as cited in F. Benes

MacLean performed a series of ablation experiments to try to determine what structures in the telencephalon might be involved in the separation call. These experiments strongly suggested that the anterior cingulate gyrus and surrounding midline limbic cortex were directly implicated in the production of the separation call (6, pp. 397–411).

Neoteny: Enhance Maternal Care: Our Highly Neotenous Brain

> "Human babies are born as embryos and embryos they remain for about the first nine months of life. If women gave birth when they should-after gestation of about a year and a half—our babies would share the standard precocial features of other primates."
>
> (23, p. 72)

The possible relationship of neoteny as a correlate to limbic elaboration in mammals was not entertained by MacLean. It is purely speculative, and is briefly discussed here only because several of the concepts in these chapters inform the discussion.

The Swiss zoologist Adolf Portmann identified two basic patterns in the reproductive strategies of mammals. It is beyond the scope of this chapter to review Portmann's work in this area; fortunately, Gould provides a succinct summary (23): Some mammals, usually designated as "primitive", have short gestation times, a large litter size, and give birth to poorly developed young (hairless, helpless, have unopened eyes and ears). The mammals exhibiting this pattern have shorter life spans, smaller brains, less complex social life. This pattern is referred to as altricial. On the other hand many "advanced" mammals have longer gestations, longer life spans, bigger brains, and engage in complex social behaviors. The young are well developed at birth. This pattern is referred to as precocial. Litter size in precocial mammals is often reduced to one (23).

Primates are the most precocial of mammals, producing very independent offspring at birth with very reduced litter size. We share most of the precocial characters with our primate cousins (long life, big brains, small litters). However our babies are born helpless, in an embryonic state—in this we most decidedly violate the precocial pattern. To account for this is it said that we are precocial mammals that give birth to babies which are secondarily altricial. Gould directs attention to the pivotal question: What is the reason for humans, the most precocial of primates, to produce at birth a helpless newborn (secondarily altricial). Gould comments, "Why did this most precocial of all species in some traits (notably the brain) evolve a baby far less developed and more helpless than that of its primate ancestors? The answer, Gould explains, relates to changes in the timing of developmental processes (heterochrony). Neoteny (holding on to youth) is a form of heterochrony, which acts to select juvenile features of ancestors and lose the previously acquired adult characteristics. Juvenile features (a potential storehouse of potential adaptations) offer such potential advantages as extended growth and development. Neotenous features, such as our highly bulbous embryonic cerebral hemisphere and cranial vault *also* result in such compromises such as our highly abbreviated gestation time. Our highly bulbous brain results in a conflict between head size and the birth canal: as one author put it the two are on a collision course. The enlarged pelvic girdle in females is not enough to ensure safe delivery, therefore gestational times are relatively shortened (the reader is referred to Gould for details). Gould comments, "Human babies are born as embryos, and embryos they remains for about the first nine months of life."

In the context of neoteny, the exuberant mammalian/higher primate limbic apparatus (also a process of neoteny) would in MacLean's theory provide necessary neurovisceral programs to provide the very extensive and extended care necessary for survival of such juvenilized young.[5]

[5] The material on neoteny is from S. Gould (23). He closes one of his essays with the couplets from Alexander Pope, which we reproduce here.

> The beast and bird their common charge attend
> The mothers nurse it, and the sires defend

CONTROVERSY SURROUNDING MACLEAN'S TRIUNE BRAIN CONSTRUCT

"... the idea that there is a distinct 'limbic system' for emotions ... is a case of a beautiful theory at the mercy of some stubborn facts"

(2, p. 24) (25, p. 576)[6]

"MacLean who introduced the term limbic system, hypothesized that the brain of mammals can be understood in terms of a tripartite organization that he has called the "triune brain." ... According to the hypothesis, the limbic system was gained in early mammals and layered over a striatal "reptile brain." ... The findings of comparative neuroanatomical investigations, as we have reviewed here, DO NOT SUPPORT THE EVOLUTIONARY HISTORY OF THE LIMBIC SYSTEM AS DESCRIBED BY THE TRIUNE BRAIN HYPOTHESIS. The present evidence indicates that THE LIMBIC SYSTEM EVOLVED LONG BEFORE THE ADVENT OF ANY AMNIOTE VERTEBRATES, LET ALONE EARLY MAMMALS."

27, p. 455–456, emphasis added.

Text book comparative neuroanatomy (27,28) acknowledges limbic infrastructure as a common element across the entire vertebrate brain series.

Limbic Components Evident Throughout the Vertebrate Series

For example, in reptiles the medial pallium elaborates two sectors, a ventromedial sector probably homologous to the mammalian dentate, and a dorsomedial sector that appears to be the homologue of the hippocampus proper. The homology is suggested by histology: these sectors in reptiles resemble the respective areas in the mammalian hippocampal formation i.e. a small cell dentate and a larger cell hippocampus anlage are identified. The homology is also suggested by a comparative study of neural

---

The young dismissed, to wander earth and air,
There stops the instinct, and there ends the care,
A longer care man's helpless kind demands,
That longer care contracts more lasting bands.

Notably, in his examination of the literature on neoteny, Gould cites workers who reported on an association of play in mammals, to large brains and slow maturation, and K selection (24). Parental (maternal) nurturing and play constitute two of MacLean's mammalian triad.

[6] Original source of this famous quotation is T. H. Huxley.

connections which reveal that a "primitive" fornix and mossy fiber system is present in non mammalian amniotes (see also figure 2.2).

All amniotes share a number of common organizational features in regard to dorsal thalamic to pallial connections. Defining features of limbic anatomy such as anterior thalamic nuclei to medial pallial districts (a canonical feature of Papez circuit) is not privileged to the higher mammalian and human brain.

Much of the information entering the medial pallial of reptiles is processed multisensory information from neighboring pallial areas. The involvement of the hippocampus (limbic pallium) and relatedly limbic subpallial structures (septal and amygdala) in higher order cerebral processes appears to be of great phylogenetic age: a shared feature of reptiles, turtles, and mammals.

Newer Insights into Limbic Brain Evolution Provide Limited Support for MacLean's Theory?

> "The elaboration of the anterior nuclear group of mammals into multiple nuclei and the shift of this system from a multisensory relay to, in effect, a cortical-thalalmo-cortical circuit (with the additional relay through the mammillary bodies) is correlated with the increased use of the medial (limbic) cortex in learning and memory. All vertebrates have ascending sensory system relays, but in mammals in particular, the analysis and use of this sensory information by the limbic system is vastly increased."
>
> (27, p. 322–323)

Butler does point to a trend in vertebrate limbic anatomy in which the midline amniote limbic district is notably enlarged compared to anamniotes; and the mammalian limbic system is more complex and expanded than that of other amniotes. Additionally, Butler draws attention to the fact that in the amniotes, and particularly the mammals, the medial pallial structures tend to be freed of channeled, sensory input from the dorsal thalamus (see below). Butler comments on the functional implications of this change in limbic afferentiation.

> ". . . the medial pallium in amniotes is relatively expanded while at the same time being freed of the task of analyzing primary sensory input. Instead the medial pallium receives so called "higher order inputs of an associative nature" and thus functions at a more complex level of information analysis. In mammals, the medial pallium is the largest and most highly differentiated relative to other amniotes, and it plays a major role

in learning, memory, and the associated functions of motivation and emotion."

(27, p. 457).

Well aware of limbic architecture in non-mammalian forms MacLean invited and responded to criticism raised. We close this section and chapter with a final comment from MacLean regarding this contended area.

"Some authors refer to the existence of a limbic system in the brain of birds and reptiles, but it is to be emphasized that the cortical area in these two classes of animals are at best rudimentary and poorly developed. Moreover, as Clark and Meyer have pointed out, structures comprising the evolutionary newest part of the limbic system (identified in the present study as the thalamocingulate division) have no representation in the reptilian brain."

(6, p. 247).

CONCLUSION

In this chapter we have examined MacLean's contribution to two aspects of limbic brain anatomy and function: 1) his efforts at expanding the confines of the limbic brain to include a dispersed limbic system; and 2) his endeavor to place the mammalian limbic brain in an evolutionary context. Only 6 years after his limbic system paper expanding the limbic domain into the frontotemporal domain Walle Nauta would publish a paper extending the limbic system into the rostral brain stem. In the next chapter we will turn our attention to the further growth of the limbic system both caudally and rostrally, with particular emphasis on Nauta's limbic midbrain.

APPENDIX

"In a sense, the three papers of Papez (1937), MacLean (1949) and Yakovlev (1948), have much in common constituting what we might now regard as a limbic trilogy. Each was concerned with defining the anatomical substrate and functional mechanisms involved in emotional expression and behavior."

(1 p. 352)

"Some delusions tend to creep into any generalization. My generalization cannot be free of them. At some risk therefore, I would like to outline the framework."

(29, p. 242)

They don't understand what it is to be awake
to be living on several planes at once

T. S. Eliot, The Family Reunion

In 1948, Paul Yakovlev published a paper, written without knowledge of Papez's work (30), which put forth a theory of limbic function which was quite similar in underlying concept to both Papez's and MacLean's formulation. In this appendix, we will briefly overview Yakovlev's general theory which includes his description of the limbic brain.

Yakovlev's theory similar to MacLean's invoked a tripartite neuroanatomic and behavioral architecture (however, Yakovlev took his concept of a trinity one step further: asserting that "the empirical representation of the entire living organism is inherently tripartite", (29, p. 242). In Yakovlev's trinity, the three germ layers of the blastocyst/body wall (von Bayer) were brought into conjunction functionally, also geometrically with the three layers of the neural tube wall (His). The reader interested in pursuing this engaging unification is referred to Yakovlev's work, several of his principal articles are in this bibliography.

Entopallium; Mesopallium; Ectopallium: Three Layers of the Hemisphere

The three outer layers of the adult hemispheric wall were termed by Yakovlev, the entopallium, mesopallium, and ectopallium, figure 3.6A. These 3 layers are arranged in concentric rings centered on an axis drawn through the lateral fissures (see chapter 1). Each pallial district constitutes a different telencephalic division: the divisions referred to respectively as the telencephalon impar, telencephalon semipar, and telencephalon totopar (see below). Yakovlev referencing classical Russian and German studies emphasized that these morphogenetic definitions did not represent idle exercises in nomenclative semantics but were justified on the grounds of evidence gathered from tectogenesis, and cyto also myelo-architectonics. Reep, in his review of the literature on the limbic brain comments. "Yakovlev's scheme is the first representation of these cortical areas based on cytoarchitectural criteria in relation to assumed function, instead of the rather haphazard equating of emotional visceral function with the grossly defined limbic lobe of Broca." (10).

The entopallium which forms a closed ring around the interventricular foramen of Monro is positioned at the keel of the hemisphere: at the hemisphere's midline hollow—the IIIrd ventricle. Yakovlev writes that it "is intimately bound to the rostral end of the brain stem, the hypothalamus—

through a felt work of unusually small, non and thinly myelinated neurons of the innermost diffuse system of visceration, (30, p. 331). The impar district and its entopallium are intimately merged with and can be regarded as the telencephalic or most rostral extension of the reticular core of the brain stem, (31, p. 255). The entopallium includes the pyriform lobe, hippocampus, and hippocampal rudiments. As Yakovlev would put it, the basic cytoarchitectonics (and relatedly wall development or tectogenetics) of entopallium represents a very limited extension (migration or elaboration) of germinal matrix. The principle activity of the entopallium, (integrated with principal activity of the inner wall structures of the body), was referred to by Yakovlev as the sphere of visceration. This sphere provided the energy to fuel the organism.

The mesopallium is the covering of an intermediate telencephalic district, the telencephalon semipar. This intermediate pallial zone forms a ring concentric to and immediately outside of the entopallium. The mesopallium covers the gyrus fornicatus, orbito-mesial wall of the frontal lobe, and the island of Reil (the outer ring structures of Broca's limbic lobe). Included in the parenchyma of the semipar is the corpora striata and its deep gray nuclear satellites, also the white matter of the centrum semiovale The semipar district functions to enable the outward expression of internal (visceral) states. It is the sphere of expressive behavior. It is similar concept to Papez's circuit, also it would seem Raymond Darts concept of mesopallial function, Dart arguing that the mesopallium district controlled the muscular display of emotion (32, p. 326).

The ectopallium invests a telencephalic district referred to as the telencephalon totopar. This outermost convexital division forms a concentric ring outside the mesopallium. Unlike the other two districts it exhibits both bilateral and spiral (helicoid) symmetry. The principle activity of this layer was called the sphere of effectuation.

Telencephalon Impar, Semipar, and Totopar

Telencephalon Impar:

This median zone can be said to contain (growing out of the lamina terminalis/reuniens) the anlage of all allocortical formations. The impar district would seem to comport with the inner ring of Broca's limbic lobe. The impar extends from the midline of the hemisphere to the limes duplex of Filimonov: the limes duplex is the border between the single layered allocortex of the subiculum hippocampus and the doubled layered subiculum. (see figure 1.9).

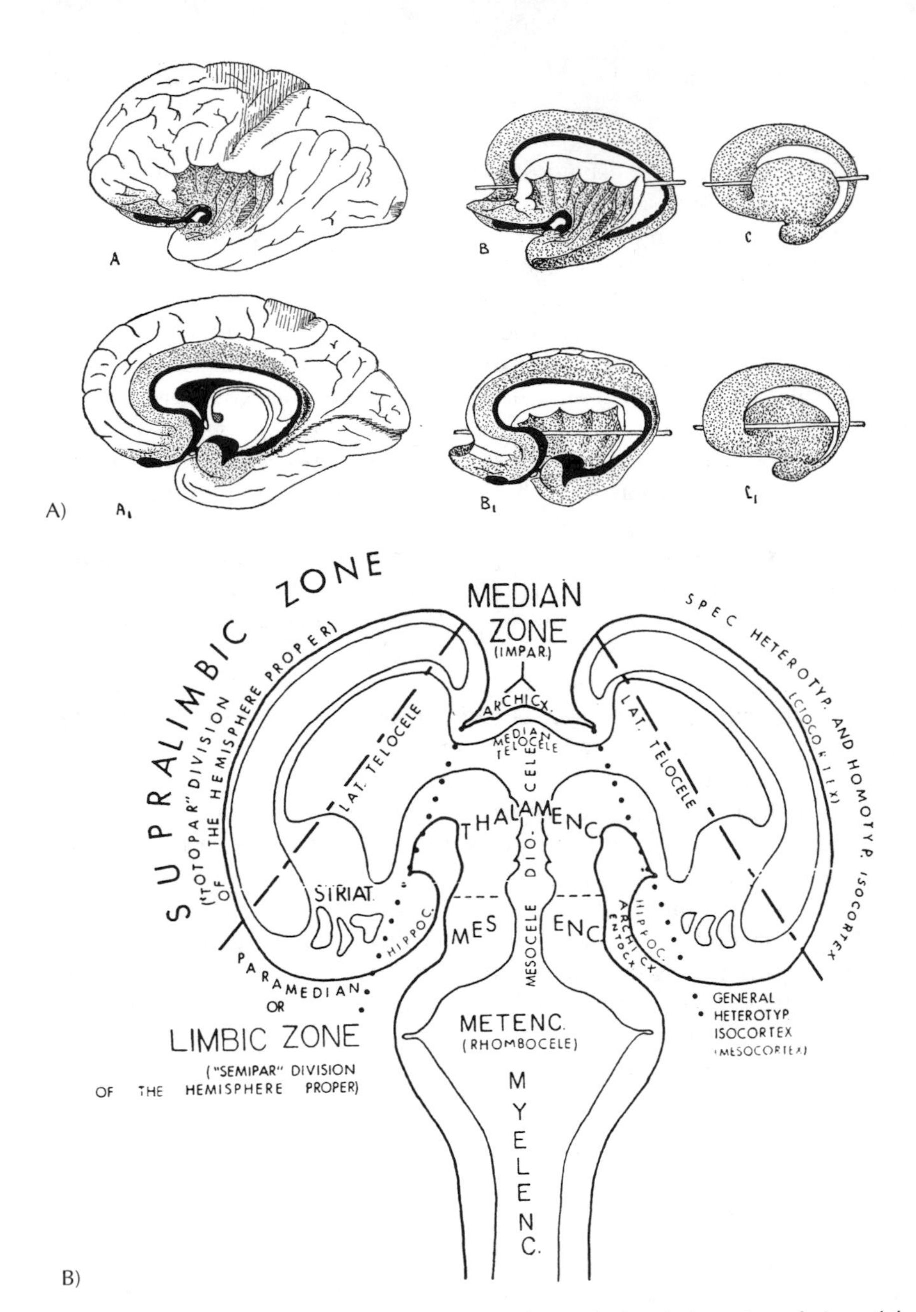

Figure 3.6. A) The three concentric rings that make up the hemisphere. Lateral view of the left hemisphere and medial view of the right. Entopallium is black; mesopallium is stippled. Epipallium surrounds the mesopallium, see text for details. B) Diagram of the three principal telencephalic divisions. The limbic zone (paramedian zone, or the semipar division of the hemisphere) mediated the motility of the body upon the body—emotion. The supralimbic zone was related to effectuation. The impar district or median zone oversaw the motility of visceration.

Semipar & Totopar

The telencephalon semipar extends from the limes duplex to the border of the true isocortex. The boundary between telencephalon semipar and totopar is gradual. The semipar district according to Yakovlev is situated proximate to the hemispheric fundal area from which the free dorsolateral and posterior walls of the hemispheric convexities emerge. It thus sits at the doorway, or limbus, to the totopar district. The semipar is isomorphic with the great expanse of Broca's lobe (gyrus fornicatus, olfactory lobe—i.e., the outer ring of Broca's lobe). Notably, in Yakovlev's icongraphy, the axes he draws to represent the semipar districts are semi parallel to each other (and thus the term semipar).

Seated over the semipar/limbic district is the totopar, "*By the end of the sixth fetal month, the supralimbic lobes assume, on each side of the interhemispheric fissure, and telencephalon impar the configuration of a pronated and cross matched pair of boxing gloves in ulnar juxtaposition over the limbic lobes.*" (31, pp. 249–252) Notably the axes Yakovlev draws for the totopar districts are, for the most part, completely parallel (totopar) to each other.

BIBLIOGRAPHY

1) Yakovlev, P. I. (1978). Limbic Mechanisms (K. E. Livingston, and O. Hornykiewicz eds.). Chapter entitled, Recollections of James Papez and Comments on the Evolution of the Limbic System Concept, Plenum.
2) Durant, J. R. (1985). The Science of Sentiment: The Problem of the Cerebral Localization of Emotion. Plenum.
3) MacLean, P. (1949). Psychosomatic disease and the "visceral brain." Recent developments bearing on the Papez theory of emotion, Psychosom. Med. 11, 338–353.
4) MacLean, P. (1952). Some psychiatric implications of physiological studies on frontotemporal portion of limbic system (visceral brain), Electroencephalography: The basal and temporal regions, Yale J. Biol. Med. 22, 407–418.
5) Papez, J. W. (1937). A proposed mechanism of emotion, Arch. Neurol. Psychiatry, 38, pp. 725–743.
6) MacLean, P. (1990). The Triune Brain in Evolution: Role in Paleocerebral Function, Plenum.
7) Nauta, W. J. H. (1962). Neural Associations of the amygdaloid complex in the monkey, Brain 85, 505–520.
8) Jackson, J. H. (1932). On affections of speech from diseases of the brain, in Selected Writings of John Hughlings Jackson, V. 2, Edited by Taylor J., London, Hodder and Stoughton.
9) Hoff, E. C. (1972). Higher cerebral regulation of autonomic function: A historical review, in Limbic System Mechanisms and Autonomic Function (C. Hockman, ed.) Charles C. Thomas.
10) Reep, R. (1984). Relationship between prefrontal and limbic cortex: A comparative anatomical review, in *Brain, Behavior, and Evolution*. 25, 5–80.

11) Cragg, B. G. (1965). Afferent connections of the allocortex, J. Anat. (Lond), 99, 339–357.
12) Scoville, W. B. (1954). The limbic lobe in man, J. of Neurosurgery, 11, pp. 64–66.
13) Hilts, P. J. (1995). Memory's Ghost: The Strange Tale of Mr. M and the Nature of Memory. Simon & Schuster.
14) Mogenson, G. J., Brudzynski, S. M., Wu, M., Yang, C. R., and Yim, C. C. Y. (1993). From motivation to action: A review of dopaminergic regulation of limbic → nucleus accumbens → ventral pallidum → pedunclulopontine nucleus circuitries involved in limbic-motor integration, in *Limbic Motor Circuits and Neuropsychiatry (*P. W. Kalivas, and C. D. Barnes eds.) CRC press.
15) Schiller, F. (1992). *Paul Broca: Explorer of the Brain*, Oxford Univ. Press.
16) MacLean, P. (1986). Culminating developments in the evolution of the limbic system: The thalamocingulate division, in The Limbic System and Clinical Disorders (B. K. Doane, and K. E. Livingston, eds.). Raven Press, New York, pp. 1–28.
17) MacLean, P. (1986). Neurobehavioral significance of the mammal-like reptiles, in*: The Ecology and Biology of Mammal-like Reptiles* (N. Hotton III, P. D. MacLean, J. J. Roth, and E. C. Roth, eds.). Smithsonian Institution Press, Washington D. C., pp. 1–21.
18) Ariens Kappers, C. U., Huber, G. C., and Crosby, E. C. (1936/67). The Comparative Anatomy of the Nervous System of Vertebrates, Including Man, three volumes, Haffner Publishing.
19) Darwin, C. (1987). Charles Darwin's Notebooks, 1836–1844: Geology, Transmutation of species, Metaphysical Enquires (P. H. Barrett, P. J. Gautrey, S. Herbert, D. Kohn, and S. Smith, eds.). Cornell University Press.
20) Angevine, J. B. (1978). Embryogenesis and phylogenesis in the limbic system, in *Limbic Mechanisms*, Plenum Press, 23–46.
21) Gould, S. J. (1993). Eight Little Piggies, W. W. Norton & Co., New York.
22) Benes, F. M. (1994). Development of the Corticolimbic System, *in Human Behavior and the Developing Brain* (G. Dawson, and K. W. Fischer, eds.). Guilford Press.
23) Gould, S. J. (1976). Ever Since Darwin, W. W. Norton & Co., New York.
24) Gould, S. J. (1977). Ontogeny and Phylogeny, The Belknap Press of Harvard University Press.
25) Shepherd, G. M. (1988). Neurobiology, Second Edition. Oxford University Press.
27) Butler, A. B., and Hodos, W. (1996). Comparative Vertebrate Neuroanatomy: Evolution and Adaptation, Wiley Liss.
28) Gloor, P. (1997). The Temporal Lobe and Limbic System. Oxford.
29) Yakovlev, P. I. (1972). A proposed definition of the limbic system, in Limbic System Mechanisms and Autonomic Function (C. H. Hockman, ed.). Charles C. Thomas, pp. 241–283.
30) Yakovlev, P. I. (1948). Motility, behavior, and the brain. Stereodynamic organization and neural coordinates of behavior, *J. Nerv. Ment Dis*. 107, pp. 313–335.
31) Yakovlev, P. I. (1968). Telencephalon "Impar", "semipar" and "totopar" (morphogenetic, tectogenetic, and architectonic definitions, International journal of Neurology 6:245–265.
32) Livingston, R. B. (1986). Epilogue: Reflections on James Wenceslas Papez, According to Four of His Colleagues (Compiled by K. E. Livingston) in The Limbic System and Clinical Disorders (B. K. Doane, and K.E. Livingston, eds.). Raven Press, New York, pp. 317–334.

CHAPTER FOUR

NAUTA'S LIMBIC MIDBRAIN

". . . the notion has developed that the unifying trait that justifies a collective name lies in the circumstance that all components of the limbic system of the cerebral hemisphere are collectively implicated in neural circuits linking the complex to a fairly well-defined subcortical neural continuum that extends from the septal region and hypothalamus into a paramedian zone of the midbrain sometimes referred to as the 'limbic-midbrain area' (Nauta, 1958).

1) NAUTA, 1972

"It cannot be denied that the hippocampal formation, together with the cingulate gyrus, the prefrontal region, and certain subcortical nuclei, are highly interconnected, and that it has been convenient to consider them, collectively as a system which is involved in functions commonly associated with the hypothalamus

2) SWANSON, p. 13, 1983

". . . it appears that the development of the meso-telencephalic dopamine system, ***which can be viewed as a component of Nauta's (1963) limbic midbrain area****, is parallel with that of the striatal and limbic forebrain regions, pointing to an intimate functional interrelationship. This also fits to the concept of a phylogenetically old interconnection between the limbic forebrain and the mesencephalic reticular formation in the so-called limbic system-midbrain circuit (Nauta, 1963)."*

3) BJORKLUND, and LINDVALL, p. 323, emphasis added

NAUTA'S LIMBIC MIDBRAIN: INTRODUCTION

Within a half dozen years of MacLean's proposal extending the limbic system rostrally to include the orbitofrontal area (4), and in the year of

Papez's death, Walle Nauta published a paper entitled, *Hippocampal Projections and Related Neural Pathways to the Mid-Brain in the Cat*, (5) which introduced the concept of the limbic midbrain and effectively extended the reach of the limbic system caudally into the paramedian midbrain. In a retrospective review addressing the principal contribution of his limbic midbrain proposal Nauta commented.

> ". . . how far into the brainstem is it possible to trace the ramifications of the limbic system? An examination of this question (Nauta, 1958) in the light of MacLean's then recent concept led to the notion that the limbic telencephalon is reciprocally connected with an uninterrupted continuum of subcortical grey matter that begins with the septum, continues from there caudalward over the preoptic regions and hypothalamus, and extends beyond the later over a paramedian zone of the mesencephalon that reaches caudally as far as the isthmus rhombencephali. The mesencephalic part of this continuum comprises the ventral tegmental area, the ventral half of the central grey substance (including the cell group now named the dorsal raphe nucleus but then known as the fountain nucleus of Sheehan), the nucleus centralis tegmenti superior of Bechterew (now better known as the median raphe nucleus), the interpeduncular nucleus, and the dorsal and ventral tegmental nuclei of Gudden."
>
> 6) NAUTA, & DOMESICK, p. 177

The body of this chapter is aimed at overviewing three related areas: In the first section we will explore the path Nauta followed in the development of his limbic midbrain construct and present a sketch of its principal features. In the following section (interconnections of the limbic midbrain circuit to frontal and spinal districts), we will, following Nauta's lead (1,6,7), briefly introduce how (consequent to the application of the histofluorescence microscope, autoradiography, also retrograde labeling of neurons with horseradish-peroxidase) core limbic structures were demonstrated to extend their reach to and enjoin more rostrally positioned districts in the neuraxis (i.e., the prefrontal cortex); also caudal districts (i.e., the intermediolateral cell column in the spinal cord). In the final section, we will consider the behavioral properties of such a highly ramified limbic subcortical axis—i.e., the behavioral properties of a limbic system embedded in the neuraxis with anatomic connections reaching the prefrontal cortex rostrally, and hindbrain, also spinal cord caudally.

THE CONCEPT OF THE LIMBIC MIDBRAIN AREA

Nauta introduced his paper by summarizing the results of prior efforts into untangling the complexity of the reticular formation. In the latter part of the nineteenth century, and early part of this one, Gudden, Cajal, Edinger and Wallenberg, had established distribution of hippocampal afflux, via the mechanism of the fornix to the basal forebrain (septal, preoptic, areas, also mammillary bodies); also the rostral midbrain gray. Valkenbug and Sanz Ibanez had demonstrated connections of the midbrain with the mammillary bodies via the mechanism of the mammillo-tegmental tract.[1] Nauta also noted that more recent investigative efforts (12,13) had demonstrated that the projection sites of the hippocampal fornix system in turn elaborated a secondary neural projection to the midbrain.

ELECTRICALLY MEDIATED ACTIVATION AND REWARD: A CHALLENGE TO MORPHOLOGISTS

"When, as we shall see, experimental studies of this part of the brain led to the conclusion that it played an important part in maintaining consciousness, what was originally a rather crude anatomical description was given a physiological meaning, exemplified by the term "ascending reticular system" which appears to mean that part of the central reticular formation which contains centripetal pathways involved in the regulation of consciousness. As these functions of the "reticular formation" were traced up into the thalamus, the thalamic reticular system became incorporated into it . . . details of the nuclei and paths in the reticular formation concerned with consciousness are still unknown . . ."

(14, p. 429)

As underscored by A. R. Luria (15), a new period in our understanding of neuroscience was ushered in by Moruzzi and Magoun when in 1949 they introduced the concept of the ascending reticular activating system (ARAS). These investigators demonstrated that electrical stimulation of the reticular formation in the regions of the midbrain, hindbrain or the lower diencephalon evoked arousal, orienting and attentional mechanisms. Moruzzi and Magoun's

[1] Investigations into the reticular formation date at least to Gall, Spurzheim, and Reil in the early 19th century (8, p. 69). In the middle of this century the ambition of the morphologist to better characterize the reticular anatomy of the rostral brain stem received a particular motivation by the observation of electrically mediated arousal, Moruzzi and Magoun (9); and the reports of electrically mediated reward, Olds and Milner (10). In a companion paper on the reticular formation published in the same year as his limbic midbrain paper (11), Nauta called attention to this issue—see side bar.

findings represented a challenge to anatomists to better characterize the particular morphology of the reticular formation, which bestowed on it such critically important functional properties (8,15)

As pointed out by Meyer, Nauta also Brain (see epigraph to this inset) despite a rich anatomic legacy of research into the reticular formation dating back to the early nineteenth century, by the middle decades of this century, knowledge of the ascending reticular anatomy (which held that ascending midbrain reticular circuits were all interrupted in the septo-preoptico-hypothalamic continuum) did not *SEEM* to endow reticular anatomy with the proper attributes to account for the phenomena of diffuse cortical arousal as demonstrated electrophysiologically. The following citation from a silver study of the ascending reticular pathways carried out by Nauta and Kuypers (11, pp. 26–27) should serve to make this issue more clear.

> *"In closing this discussion, it is necessary to point out that none of the ascending reticular projections outlined in the present study can offer an obvious explanation for the phenomenon of diffuse cortical arousal so clearly demonstrated by physiologic experimentation. Indeed, it may be logical to ask: Do any of the pathways here traced actually ascend beyond the confines of the "reticular formation"? It is obvious that the answer to this question will depend on largely personal and arbitrary criteria, but reasons could be found to include the hypothalamus, septal region and perhaps even the intrathalamic cell groups, habenula and reticular nucleus of the thalamus in the "reticular formation". Ascending reticular projections can be traced to all of these structures and, in addition, to the basal ganglia. It must, however, be emphasized that at the present time no structures outside the "specific" thalamic nuclei of the thalamus have been definitively demonstrated to project significantly to the neocortex and the role of these nuclei in the mechanisms of diffuse cortical activation still appears ill understood. Thus it must remain for further studies to identify beyond hypotheses the final corticipetal link in the anatomical substratum of the phenomenon of diffuse cortical arousal, the discovery of which perhaps more than that of any other functional manifestation, is to be credited for the wide spread of current interest in the reticular formation."*

A possible anatomical explanation for ARAS, pointing to direct innervation of the cortex by the reticular activating system, would be realized, but not with silver methods. At the same time Nauta was establishing the limbic midbrain circuit in silver studies (5,11) researchers at the Universities of Gotesborg and Lund in Sweden were adapting the fluorescence microscope for the examination of monoaminergic cells in the central nervous system. It fell to fluorescence microscope to "identify beyond hypothesis" the final corticopetal links in the anatomical substratum: a link that might yield a more satisfying explanation to the electrophysiologically characterized phenomenon of diffuse cortical arousal.

Nauta did not appear to rule out multineuronal reticulo-thalamic, also reticulo-subthalamic pathways as mediating the phenomena of ARAS (5, p. 332); but he seems to suggest from the perspective of a mid-century authority on reticular anatomy that direct pathways, if they were to be found, would provide a much more likely mechanism (see above). Do the ascending monoamine fibers represent the mechanism of direct cortical arousal? In this regard, it should be noted, that in his review of the MFB written in 1982, Nieuwenhuys, remarks that it is not clear that the phenomenal of ARAS is correlated with direct monoaminergic (dopaminergic) fibers to the cortex (16).

Generalizing on these findings, Nauta, entertained the idea of a neural network with convergence areas or nodes rostrally situated in the basal forebrain ("*Most . . . such projections have their cranial representation in palaeocortical and subcortical structures in the basal and medial walls of the cerebral hemisphere, collectively designated by the term limbic system.*" (5, p. 320) and caudally in the paramedian midbrain. The neural network served to pass, in a reciprocal fashion, neural code from such structures as the hippocampus and the amygdala to the midbrain (over such fiber pathways as the medial forebrain bundle). *Limbic* would be the descriptor for this pathway—see text below. Nauta's hippocampal projection paper represented his attempt to subject the generality of this notion to further experimental test.

Nauta's Limbic Midbrain: Methods; Finding; Pathways

> "At this point I would like to pay a tribute to Dr. Nauta as he is here with us. His technique is probably the most sensitive and reliable method in use for tracing connections of the central nervous system today. During the last twenty years it has contributed enormously, and I am sure that in the history of experimental neurology his name will rank with others of the past such as Nissl, Golgi, and Marchi."
>
> (17)

A brief statement regarding Nauta's method is in order. Selective electrolytic lesions were made in the lateral preoptico-hypothalamic area, and employing a modification of silver staining (i.e. his own suppressive Nauta-Gygax method) the resultant neuronal degeneration was traced. This method allowed the selective silver impregnation of degenerated axons and axon terminals, while excluding the staining of normal fibers, see figure 4.1. Fine-scaled studies of brain stem connectivity's could not have been done successfully without employing such modifications of silver staining. Lennart Heimer commenting on the historical importance of the Nauta-Gygax technique remarked that the development of the so-called suppressive Nauta-Gygax method, which afforded a selective suppression of the argyophilia of normal axons marked the beginning of a new chapter in the history of silver impregnation (18). The particulars of Nauta's finding are beyond the space limitation of this text. Nauta provides a summary statement

> "Indirect hippocampal pathways, in part interrupted by further relays, and probably representing indirect projections of the amygdaloid complex, originate in the septum and in the lateral preoptic and hypothalamic regions, as well as in the mammillary body, Such pathways reach the mid-

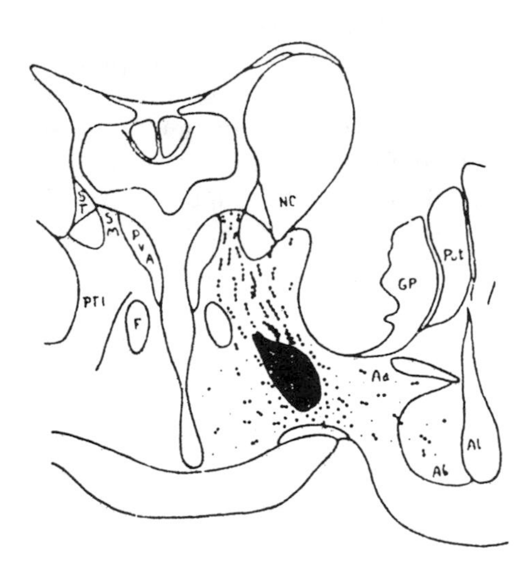

Figure 4.1. This is from the first figure in Nauta's hippocampal projections paper. Fiber degeneration is shown following a lesion in the lateral preoptic area. The electrolytic lesion is indicated in jet-black; terminal fiber degeneration in fine stipple; fibers of passage, coarse stipple. This study also included lesioning more caudally in the lateral hypothalamus, and also placement of a lesion in the fasciculus retroflexus. Figure reprinted with permission, from (5).

> brain in three fibers systems, viz., the medial fore-brain bundle, the fasciculus retroflexus, and the mammillo-tegmental tract."
>
> (5, p. 339)

Having more firmly established the presence of anatomical interrelationships of the midbrain area with structures of the basal forebrain, traditionally considered part of the limbic brain, Nauta pushed the limbic designator caudally.

> "Since . . . it [the paramedian midbrain] appears to be related specifically to the limbic system, it seems justifiable to use the term **limbic mid-brain** area as a schematic designation of the regions here discussed (authors emphasis)."
>
> (5, pp. 333–334)

In the following subsections, three main pathways providing neural conduit in Nauta's limbic midbrain construct are overviewed, see figure 4.2 for a schematic which demonstrates how these pathways "fit in" with the hemispheres white matter cabling infrastructure.

The Medial Forebrain Bundle (MFB)

Nauta points out the MFB was named by Ludwig Edinger to contrast this fiber pathway with the far more massive fiber system which Edinger called the lateral forebrain bundle (internal capsule and peduncle). At the level of the diencephalon, the MFB is located in the lateral hypothalamus, see figure 4.3. Nauta underscores an important principle, evocative of the hemispheric dynamic that informed Herrick and Papez, relevant to the "limbic location" of the MFB.

> "The medial forebrain bundle is in essence a limbic traffic artery; its medial placement suits the medial placement of the limbic structures it serves. The lateral forebrain bundle is in essence a neocortical traffic artery; its lateral placement suits the lateral placement of most of the neocortex."
>
> p. 265, (20)

As characterized in his hippocampal projections paper, the rostral extent of the MFB obtained the hypothalamico-septal continuum, and the caudal extent engaged the paramedian midbrain. The MFB is interpersed throughout its extent with gray matter in an arrangement that is

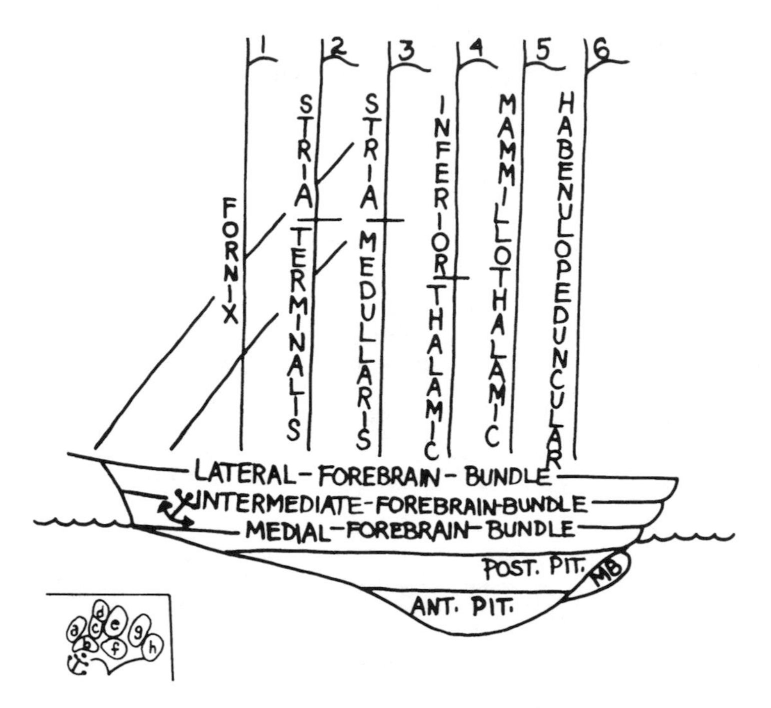

Figure 4.2. Scheme for representing the main neural pathways of the forebrain. The medial, lateral (also intermediate) forebrain bundle form principle horizontal pathways in the basal forebrain. Six pathways are depicted which have a more or less vertically oriented direction: of this vertical consortium, the fornix is the most anteriorly positioned, and the habenulopeduncular tract the most posteriorly positioned. In MacLean's schematic, the impression of a sailing ship is intended. Four of these "vertical" pathways, the stria medullaris, mammillothalamic tract, habenulopeduncular pathway, and the inferior thalamic peduncle, constitute principal conduction pathways in Nauta's limbic midbrain circuit. Nauta's limbic midbrain circuit paper considers these the secondary pathways (5, p. 319) acting to shuttle (and expand) neural traffic between limbic junctional areas in the lateral preoptic and hypothalamic region and the midbrain. In this regard the fornix, and stria terminalis (discussed in the first chapter) would be considered the primary pathways. Figure from MacLean (19).

characteristic of the brainstem reticular formation (thus its often vague delineation, and the difficulties it presents to visualization by fiber degeneration methodologies). Embedded in the parenchyma of the MFB are clusters of gray matter (viz. interstitial nuclei of the MFB). Tsai's ventral tegmental area (the term haube, also tegmentum refer to neural structure roofing, or forming a shell or husk around other neural structures), Bechterew' nucleus (the median raphe), Gudden's tegmental nuclei, and

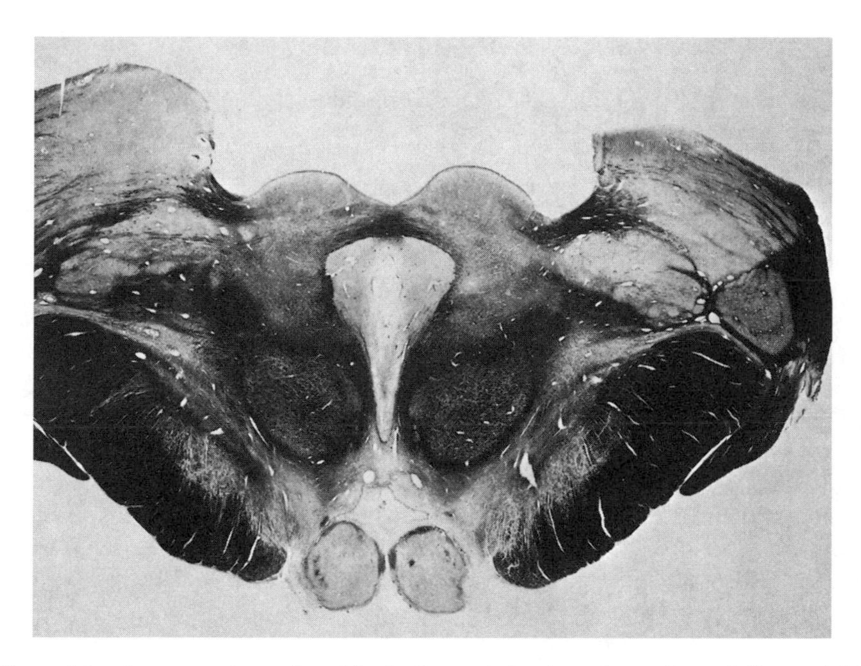

Figure 4.3. A cross section at the midbrain diencephalon (aqueduct—diencephalic ventricle) transition. The area between the red nucleus (the red nuclei are the spherical loci located 3–4 cm directly dorsal to the mammillary bodies, they stain less darkly in this section stained for white matter) and the interpeduncular fossa is part of the lateral hypothalamus. This area is called the ventral tegmental area of Tsai, and is proximate to an area formerly called Papez's nucleus of the mammillary peduncle). The MFB tracts through this area. The substantia nigra is seen as pale nuclear structures seated atop the peduncles (SNc and SNr are not discriminated in this diagram). Figure from Nauta, legend adapted (20).

the locus coeruleus, represent interstitial nuclei clustered toward the *caudal* end of the MFB: the lateral preoptic nucleus and other hypothalamic nuclei are interstitial nuclei located at the *rostral* end of the MFB.[2]

[2] Three additional aspects of the MFB are worth mentioning. 1) The MFB, although a prototype for reticular traffic, ALSO carries information from the classical sensory spinothalalmic pathways (either by synaptic relay, or by accepting fibers from these classical pathways). 2) Nauta cites an earlier article by Tello (1936) which suggests that the MFB may precede the fornix in development (5, p. 324). This finding suggests that the MFB may represent one of the earliest limbic cabling systems to be laid down during neuroembyrogenesis. 3) As underscored by Nauta in his textbook, "*. . . the medial forebrain bundle has emerged in recent years as the great highway for monoaminergic fibers ascending from the hindbrain and midbrain into the forebrain. It is the only place in the brain where the main conduction lines for the three monoamines commingle.*" (20, p. 126). For a rather exhaustive treatment of the MFB see (16).

The Stria Medullaris (SM) and the Fasciculus Retroflexus (FR)

As noted above, the SM (septal to habenular nuclear fiber pathway, this neural conduit, forms one of the three anteriorly positioned "vertical" pathways in figure 3.2), and the FR (habenular to interpeduncular nucleus neural channel) played an important role in Nauta's formulation. Lesioning the lateral preoptic area, see figure 4.1, Nauta tracked fibers of the stria medullaris as they gained the habenular nucleus. Then he lesioned the FR as it exited the habenular (figure 14.20(s) depects this lesion and subsequent degeneration). The course of the FR was then traced to the basal surface of the midbrain.

The habenular, its major efferent pathway the FR, and the interpeduncular nucleus (a principal terminus for the FR), are common features of all vertebrates and are assumed to have been present in the ancestral stock of all vertebrates (21). This conduction system provides an important route for frontal olfactory and limbic loci to access the midbrain tegmentum and therefore enables the frontoseptal and olfactory information to be integrated with spinal and brainstem sensory information. Reciprocal projections of the interpeduncular nucleus → hippocampus → fornix → septal area → habenular effectively place the habenula as a node in another limbic circuit.

Mammillotegmental Tract

The mammillotegmental tract was a third fiber system forming a principal element of Nauta's construct. According to Meyer, Cajal was the first to demonstrate with undisputed clarity that the mammillothalamic and the mammillotegmental tract arose in the medial nucleus of the mammillary body from a common stem (8, p. 62). This common stem is said to move anterodorsally, and then bifurcates into the mammillothalamic tract (part of Papez's circuit) and the descending mammillotegmental tract. Affirming earlier studies, Nauta lesioned this area and traced the mammillotegmental tract to the caudal midbrain where it enters Gudden's deep tegmental nucleus.

SUMMARY TO SECTION

With the publication of Nauta's, hippocampal projections paper, limbic anatomy embraced a caudal domain not included in the original descriptions of Broca, Papez, or MacLean. As the concept stood in 1958, Nauta's limbic midbrain circuit was more or less closed and bounded by hypothal-

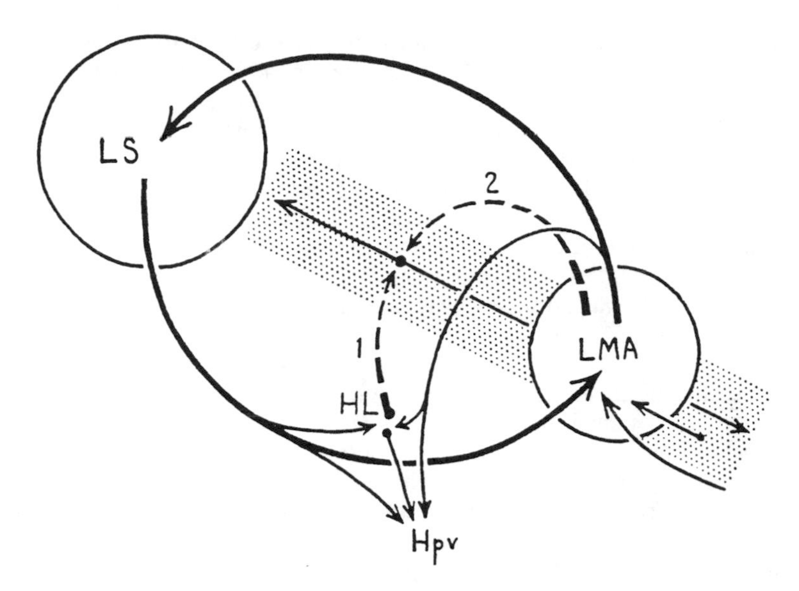

Figure 4.4. Simplified diagram of the limbic system midbrain circuit. The darker arrows indicate the main pathway of the bidrectional limbic midbrain circuit. Curved broken arrows indicate collateral, and escape pathways. Ascending pathways also partaking in Nauta's limbic midbrain circuit included: 1) the dorsal longitudinal fasciculus, a neural conduit between the central gray of the midbrain, and the hypothalamic septal continuum; 2) fibers collecting in the mammillary peduncle and continuing rostralward in the MFB, also distributing to the hypothalamic and septal areas; also 3) lateral and medially positioned components of the brainstem reticular formation conjoined with somatic sensory (trigeminothalamic, spinothalamic) components which reached subthalamic and thalamic levels. The presence of the ascending limb of the limbic midbrain circuit prompted Nauta, as early as 1958, to consider a role for this circuit in the activation of endocrine and autonomic mechanisms also affective mechanisms. Figure reprinted with permission from (5).

amic and hippocampal centers rostrally, and paramedian loci caudally however, many escape routes were noted, see figure 4.4. It is important to note that even at the time of his original publication, Nauta entertained additional connections to his limbic midbrain circuit originating as far rostrally as the neocortex.

INTERCONNECTIONS OF LIMBIC MIDBRAIN CIRCUIT TO SPINAL AND FRONTAL (NEOCORTICAL) DISTRICTS

"Although it is well known that the limbic system exerts a strong influence on somatic and autonomic motorneurons, lesion-degeneration studies did not reveal strong limbic projections to levels caudal to the mesencephalon. This led to the idea that the limbic pathways to caudal brainstem and spinal cord were multisynaptic (Nauta, 1958). Since 1975 this view changed dramatically mainly because new tracing techniques became available such as retrograde tracing using horseradish peroxidase (HRP), anterograde autoradiographic and immuno-histochemical fiber-tracing techniques. Kuypers and Maisky (1975) using the retrograde HRP technique demonstrated direct hypothalamico-spinal pathways in the cat. Subsequently, the autoradiographic tracing technique has revealed many new limbic system pathways to caudal brainstem and spinal cord."

22, p. 369

"The mesencephalic dopaminergic systems form major components of the catecholamine innervation of the neocortex. As such they contribute to the non-specific afferent innervation of cortical neurons and has been suggested by others may have an important role in neocortical functions. **From the purely neuroanatomic viewpoint the neocortical dopamine innervation represents a direct projection from midbrain to telencephalon which has not been previously described.**"

23, p. 330, emphasis added

Nearly 25 years after its initial publication, Nauta reexamined his original construct in light of research findings employing newer tract tracing technology (6). As in previous chapters, we will turn to the author for a summary statement of how these findings impacted on his earlier contribution. (A review of these findings is a rather studied exercise beyond the scope of this text, see, 6,16,22).

"It is plain from the foregoing survey that recent anatomical analyses have added much detail to the previously elaborated picture of subcortical limbic connections. None of these newer findings, however, appears to contradict the initial impression that these connections in part compose a circuit reciprocally linking the limbic structures of the olfactory tubercle and septal region caudalward over the preoptic region, the hypothalamus and the paramedian zone of the midbrain. The most important revision of the original construct is dictated by the evidence that the ascending limb of the circuit is composed in part at least of fibers, especially of the monoamine variety, that extend directly to the limbic structures of the cerebral hemisphere. In the initial report, the ascending return loop of the circuit was assumed to be quantitatively interrupted in the septo-preoptico-hypothalamic continuum."

(6, p. 198)

Nauta acknowledged that his limbic circuit was embedded in a larger more ambitious domain of limbic affiliated structures. However, he underscored the point that this newer perspective could be seen as subsuming and expanding, not contradicting, his original proposal. Two aspects of this limbic elaboration can be summarized as follows: 1) Connections interfaced with the limbic midbrain circuit extended rostrally to frontal areas caudally to the cervical cord. 2) Consequent to fluorescence microscopic examination of the rostral brain stem, it was revealed, for the first time, that the paramedian midbrain (a critical convergence point in Nauta's original construct) was a principal component of a thalamic bypass circuit. The bypass circuit meant that the paramedian midbrain could obtain the limbic cortex *without* synapse at the thalamus.[3]

It should be said that the demonstration that limbic structures such as the hypothalamus amygdala, and related forebrain structures innervate a) branchial motor centers (the Vth and VIIth cranial nerves: the centers for facial expression); b) visceral afferent centers (the solitary nucleus); c) the autonomic column, and that fact that core limbic structures engaged rostral loci, such as the prefrontal region (this issue is discussed below) served to establish a close integration between central and peripheral determinants of emotional behavior. Anatomic integration of this degree, a long sought after goal of such pioneers of the limbic brain as Papez and MacLean, points the way to a resolution of old disputes between such views as the James-Lange theory (peripherally weighted theory of emotion) with more centrencephalic determined theories.[4]

[3] Afferentiation of the cortex, without a synapse at the thalamus (save olfaction, and to a limited extend gustatory sensation) was a radical departure from conventional thinking at mid century. In the wake of the histofluorescence microscope, the previous anatomic wisdom, which held that projections to the cortex had to be relayed first through the thalamus, was no longer tenable. The question is raised, what, if any, implications for behavioral medicine and psychopharmacology are there stemming from monoaminergic fiber pathways obtaining the cortex directly.

[4] The reader is referred to a summary of the James-Lange position by Walter Cannon (24, p. 52). Cannon underscores the fact that in the James—Lange theory afferent impulses from disturbed peripheral organs contribute to (indeed largely determine) the emotional state. Pietro Corsi provides a succinct statement of the James-Lange position, *"In other words, there Is no primary 'seat' of emotion in the brain, there is only neurological feedback from those visceral organs that are involved in the different physiological states we experience as sorrow, anger, joy, etc."* (25, p. 214). It would appear, in light of the intimate integration of peripheral anatomical components with central components, that peripheral (James-Lange) versus central theories (Herrick, Cannon, Papez, MacLean) are two sides of the same coin.

The Use of Newer Methods Demonstrate Direct Afferentiation of Limbic Cortex from the Paramedian Midbrain: Fluorescence Microscopic Mapping of the Ascending Monoamine System

In the following subsections, the impact of the fluorescence microscope on Nauta's limbic midbrain is overviewed[5]. By intent, and by constraints of space, this subsection will highlight only select nuclei from the paramedian midbrain. Also only one ascending monoaminergic tract, the mesolimbic tract, (section B–1C below), will be discussed because of its limbic designator.

Technical Considerations

Serotonin, also dopamine and norepinephrine combine with formaldehyde (as a vapor) to yield a compound that fluoresces at characteristic wavelengths under ultraviolet light. This is the basis for the cellular localization and mapping of the monoamines in the fluorescence microscope[6].

Nuclei

Employing the histofluorescence microscope, Anica Dahlstrom and Kjell Fuxe proposed a scheme for the identification of the various monoamine nuclei in the brainstem (27). In their systematic survey, the catecholamine nuclei were all considered part of the (A) group (recall that DA and NE; also epinephrine are catechols by virtue of the presence of the catechol, or dihydroxy phenol group in their structure). The serotonin

[5] It should be appreciated that the application of the histofluorescence microscope to tracing monoaminergic tracts in the rostral brain stem was not motivated to extend Nauta's limbic midbrain construct—this is hardly the case. The development of histofluorescence microscopy and its application to the nervous system was closely linked to research focused on monoamines in the rostral brain stem, also research on Parkinson's disease—research which pointed to the role of dopamine as a neurotransmitter in the brain, and suggested the probability of dopamine projection systems in the rostral brain stem. For historical overview of this material, see (26).

[6] Nauta emphasizes the point that the Falck-Hillarp histofluorescence technique by enabling direct visualization of the monoamine neuron and avoiding such interventions as axon transection, behaved like a true stain revealing the distribution of something inherent in the brain. The histofluorescence technique is capable of demonstrating the existence of monoamines in cell bodies, axons and axon terminals, and can consequently map out individual nuclei and neural pathways for each neurotransmitter; however, it should be noted that because midaxonal concentrations of the monoamines are often too low to visualize, pharmacological manipulation and lesioning is often necessary to better visualize pathways.

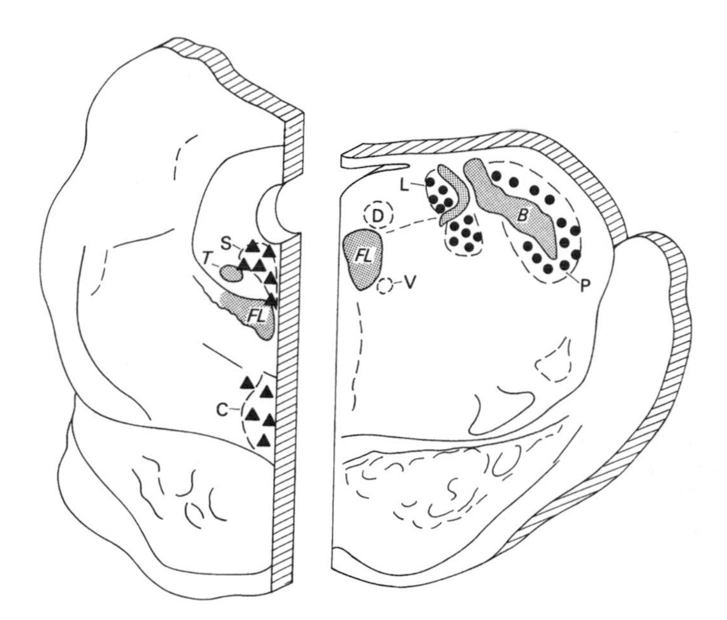

Figure 4.5. Two cross-sections taken at the isthmus\pontomidbrain transition. The section at the left is at the extreme caudal border of the midbrain. The section on the right, a bit more caudal, is positioned in the rostral pons (the IVth ventricle betrays this more caudal section as pons). Nuclei of particular interest are found in these sections (trochlear, supra trochlear, dorsal and median raphe locus coerulus (see MacLean (19) for details). Interestingly, many of these pigmented nuclei were identified with the regular light microscope on the basis of intraneuronal neuromelanin (a catecholamine condensation product). A condensation product of monoamines (with formaldehyde vapor) would also be responsible for their imaging with the fluorescence microscope. Figure reproduced with permission from (19).

nuclei were termed the (B) group. The maps of these monoaminergic pathways in man and in rodents are available in many basic chemical neurobiology texts and will not be reproduced here, only a brief description of the nuclei follow.

Catecholaminergic Group (A1–A7): In their original paper the authors only identified 12 catecholamine groups (A1–A12). The A1–A7 cell groups are noradrenergic. All of the noradrenergic nuclei are located in the hindbrain. Their fibers ascend in the tegmental bundles and join the MFB in the rostral brain stem and forebrain. Cell group A6 is the locus coerulus. The locus coerulus labeled "L" in figure 4.5 is partially roofed over by a root of the Vth nucleus.

Catecholaminergic Group (A8–A12): A8–A16 cell groups are dopaminergic. Dopaminergic nuclei *are not found* in the hindbrain. A8–A10 are found in the mesencephalon. A11–A13 are positioned in the diencephalon. A16 is located in the olfactory area (telencephalic). It is interesting to consider what evolutionary significance may be attached to the caudal to rostral concentration gradient for NE and DA.[7] The A10 neuronal group corresponds to the VTA of Tsai. It is located just medial to the substantia nigra. The A10 designator serves to discriminate this area from the substantia nigra pars compacta which was designated cell group A9. Cell group A8 is encountered in the retrorubral area, lateral to the nigra. The diencephalic group A12 is located in the median eminence. This nucleus contributes to the hypothalamopituitary portal system.

Serotoninergic Group: B1–B9 are serotoninergic. B1–B4, which constitute the caudal group, are found is located in the hindbrain and upper cervical cord. B6–B9 are considered the rostral brain stem group (B5 straddles both groups). Within the rostral brainstem group, B7, B8, and B5, constitute the dorsal and median raphe. The rostral group gives rise to the ascending serotoninergic projections to the forebrain. The word raphe is applied to many of the serotonin nuclei because of their distribution close to the midline; however, in many forms, particularly higher mammals, also birds, there is a lateral spread of serotonin cell clusters in the brain stem; this results in an intermingling of catecholamine and serotonin cells in the brain stem, the MFB, also in the forebrain.

Ascending Tracts

Mesolimbic: Of all the ascending monoamine fiber systems, one dopaminergic system appeared (to early histofluorescence microscopists) to so evidently display its limbic affinities that it was called the mesolimbic tract. We examine this tract to study yet another example of the application of the limbic moniker.

As mentioned above, the label A10 was given by Dahlstrom and Fuxe (27) to the large wedge shaped DA cell cluster located just medial to the substantia nigra. Shortly thereafter Anden (29) reported histochemical findings combined with biochemical evidence that enabled him to subdivide the ascending DA fibers from the midbrain into —) a nigrostratal

[7] The differentiation of the catecholamine family suggests a molecular example of parcellation theory. Parcellation theory as a mechanism for evolution of the nervous system argues that pre-existing structures differentiate through splitting off (parcellation). Parcellation theory avoids the need to invoke *de novo* creation of new structure, and it provides a mechanism for homology: in parcellation theory a new nucleus does not appear all of a sudden but daughter (satellite) nuclei evolve from a common structure. See (28, pp. 22–23) for a brief introduction to this issue.

system in which fibers originated from the pars compacta of the substantia nigra (A9) and innervated striatal zones (caudate and putamen),[8] and —) an ascending system originating from cells medial to the nigra, the VTA A10 cell group, and projecting to the basomedial region (ventral sector) of the striatum. Subsequently, Ungerstedt (30) determined that the A10 neurons also sent projections to the central nucleus of the amygdala, also the bed nucleus of the stria terminalis (a major amygdala outflow pathway). On the basis of these findings (i.e., the amygdala, at least since MacLean's paper, being a major limbic locus), Ungerstedt introduced the term mesolimbic system to distinguish these ascending projections of DA cell group A10 form the nigrostriatal projections originating from A9.

Nauta marshals further evidence to legitimize the limbic designator for this tract (31). Again the issue here is not to assert that the mesolimbic tract as canonically limbic, but to review some historical issues attached to its name.

- the mesolimbic system projects to that part of the striatum that receives its descending (telencephalic) afferents from limbic centers (hippocampal and amygdala) as opposed to that part of the striatum that receives mainly neocortical input.
- several papers published in the 1970's demonstrated that the mesolimbic system projects to areas in the frontomedial cortex, which also receives a heavy afflux from the anterior and medial thalamic nuclei. The anterior thalamic nuclear complex played a central role in the limbic concept of Papez. The convergence of A10 projections and limbic thalamic projections on common frontal cortical areas provided added support to legitimize the limbic affiliation of the mesolimbic pathway.
- the A10 VTA cell group is imbedded in the interstices of the MFB—it therefore lies embedded in descending fibers originating from basal forebrain limbic centers.

---

[8] In the narrow sense of the term nigrostriatal tract refers to the A9, pars compacta DA neurons whose axons project to the dorsal aspect of the caudate and putamen. This is the most massive of the efferent nigral connections. If one considers the A10 VTA projection, a medial nigral efferent, then the DA fiber projections from the nigra are yet more extensive and cover the whole of the caudate and putamen. The A10 projections are restricted largely to the ventral sector extending up to midstriatal levels where they overlap with A9 projections. The extreme dorsolateral sector of the striatum, rich in A9 projections, does not appear to receive any A10 projections. Conversely there are few A9 projections in the most ventral striatal districts. An examination of the ascending nigrostriatal connections suggests the concept of two striatal districts: a dorsal sector rich in A9 innervation, and a ventral region preferentially in receipt of A10 afflux. The concept of a ventral striatum will be introduced in chapter 5.

The Mesocortical Dopamine Tract: As noted above, during the early 1970's, knowledge of the midbrain dopaminergic fibers was further expanded allowing the identification of midbrain to cortical DA systems—specifically DA projections to the cingulate, entorhinal, suprarhinal, piriform, and frontal cortex (32). To describe these projections the term mesocortical dopamine system was coined.

The Mesotelencephalic System: Ascending DA projection systems were referred to as nigrostriatal, mesolimbic, and mesocortical (32). In light of the considerable overlap, and certain terminological inconsistencies between these systems, Bjorklund and Lindvall argued for the concept of a mesotelencephalic system to refer to the entire ascending forebrain projections of the midbrain DA system. Two major subsystems are distinguished: the meso-striatal (including all DA projections to the caudate-putamen and accumbens), and the mesocortical which would include all cortical projections from intrinsic midbrain DA nuclei.

AN EXTENDED LIMBIC SYSTEM: NEUROBEHAVIORAL IMPLICATIONS

The material covered earlier in this chapter argues for a rather extended limbic embodiment. A consistent topic addressed by Nauta dating to his hippocampal projections paper, was the behavioral implications of such an extended limbic coalition with particular reference to the extension (connections) of the limbic continuum with frontal districts (1,5,6,7,20,31). There is a rather extensive body of anatomical literature on the discovery of frontolimbic connections (we limited our discussion to an overview of monoamine circuits between the midbrain and frontal regions). Nauta's commentary below, although summary, does render a sense of both the anatomy and the behavioral anatomic implications of the extensive frontolimbic interrelationships

> "Nowhere more remarkably than in its subcortical connections does the frontal lobe declare its close association with the limbic system. This relationship is most explicitly expressed by the substantial projections that have been traced from the frontal cortex to the preoptic region and hypothalamus and, beyond the diencephalic structures, to a paramedian zone of the mesencephalic tegmentum . . .
>
> The frontal lobe is characterized so distinctly by its multiple associations with the limbic system, and in particular by its direct connections with the hypothalamus, that it would seem justified to view the frontal cortex as the major—although not the only—neocortical representative of the limbic

system. The reciprocity in the anatomical relationship suggests that the frontal cortex both monitors and modulates limbic mechanisms."

33, pp. 180–182

Some of the behavioral issues evoked by the concept of limbic-frontal circuitry are briefly sketched below.

— *Bi-directional Neural Traffic (Cross Talk) Between the Frontal and Basal Limbic Districts*: Nauta submits that the interaction of the frontal cortex and the limbico-subcortical axis "could be, among other things, a prerequisite for the normal human ability to compare alternatives of thought and action plans." Nauta continues, "*This suggestion attributes to the limbico-subcortical axis the function of a 'sounding board' or 'internal test-ground' enabling man to preview the affective consequences of any particular action height consider, and thus, to permit him a choice between alternatives of thought and action*.". Notably as pointed out by Kuhlenbeck (34) the concept of reverberating neural circuits elaborating awareness behavior does not appear to have received emphasis prior to Papez's proposal relating emotion to cortico-subcortical circuitry. Nauta's ideas regarding frontolimbic reentrant circuit acting as a sounding board elaboration our inner voice seems to echo Papez's central idea.
— *Loss of Frontal Activity (Decreased Corticofugal Activity)*: The loss of frontal neocortical activity on intrinsic (basal hemispheric) limbic systems is hard to quantify; however, it is certain to eliminate important pathways by which the neocortex modulates the organism's affective and motivations states. Frontal lobe dysfunctions commonly present a wide compass of clinical findings from failure of affective and motivation responses to appropriately match environmental contingencies to marked abulic, amotivational states
— *Loss of Rostral Brain Stem Basal Hemispheric Limbic Activity (Decreased Corticopetal Innervation)*: Amotivational, anergic, depressive states are also associated with subcortical dysfunction probably involving compromise of ascending monoaminergic influence.

Clinical Localization in Psychiatry and Neurology: Frontolimbic Injury in the Case of Phineas Gage

During the last four decades of the nineteenth century (a period at the very beginning of the belle époque of neurology (40)) the principle localizations providing the basis for clinical neurology were first described (principle locations included Broca's description of motor/anterior aphasia in 1861, Wernicke's description of posterior/sensory aphasia in 1874, Hermann Munk's work suggesting an occipital cortical localization for vision in 1881). Clinical pathological correlation as the operative principle for clinical neurology was born on the lathe of localization: where is the lesion localized, and what is the nature of the lesion are canonical to every clinical neurology conference dating from this era. It has been argued that clinical psychiatry parted company with clinical neurology during this era; what could not be localized was jettisoned off into clinical psychiatry (38,39,40).

Heralding this era, and heralding the divide between the two clinical sciences, was Broca's presentation of Mr. Lebourgne (alias Tan Tan—the alias recognizing the one syllable the patient could utter) to the anthropological society of Paris in 1861, the year of Lebourgne's death. In this last section of this chapter, we submit for consideration, the case of Phineas Gage, who also died in 1861 (although Dr. Harlow's autopsy report of the skull did not appear until 1868) as an example, at least in part, of frontolimbic dysfunction presenting with neuropsychiatric disturbance. The argument receives some support from the remarkable reconstruction of Gage's injury which shows the pathway of the projectile injury (Gage's skull, and brain was penetrated by a tamping iron)—the projectile passed through frontolimbic districts (35,36,37). Notably, although some would object that Gage's case involved rather extensive destruction of frontal areas, Broca's case of anterior aphasia was also confounded by rather extensive neuropathological changes reaching well outside so-called Broca's area.

Our point in this concluding section is to submit an argument (based on features of frontolimbic anatomy) for chronologic parity in localization for both clinical psychiatry and clinical neurology, and to do more; to point out that the division between both fields is and always has been a false dichotomy. We end this chapter with a quotation from Samuel Kinnier Wilson:

> "The antithesis between 'organic' and 'functional' disease states still lingers at the bedside and in medical literature, though it is transparently false and has been abandoned long since by all contemplative minds."
>
> KINNIER WILSON (41)

CONCLUSION

In this chapter we followed the extension of limbic connections to the paramedian midbrain, as put forth in Nauta's proposal, section A, and even more caudal districts as characterized by subsequent anatomical research Limbic connections with frontal areas were also considered, as were the behavioral correlates of such connections. Disturbance of the frontolimbic axis (in the case of Phineas Gage, see last inset, argues for historical parity for localization of emotional (psychiatric) and language disturbance. It should be noted that several other well-regarded neurobehavioral theories posit a construct similar to Nauta's septo-hypothalamic, e.g. Penfield's centrencephalic core (1952); also Luria's first functional unit (1973). In the final chapter of this book, we will examine one further anatomic interrelationship further extending the compass of limbic anatomy: the concept of the limbic striatum.

REFERENCES

1) Nauta, W. J. H. (1972). Connections of the frontal lobe with the limbic system, in Surgical Approaches in Psychiatry (L. V. Laitinen, and K. E. Livingston, eds.). University Park Press, Baltimore, pp. 303–304, *Proceedings of the Third International Congress of Psychosurgery*, University Park Press Baltimore.
2) Swanson, L. W. (1983). The hippocampus and the concept of the limbic system, *in Neurobiology of the hippocampus* (W. Seifert, ed.). London Academic Press.
3) Bjorklund, A., and Lindvall, O. (1978). The meso-telencephalic dopamine neuron system: A review of its anatomy, in. (K. E. Livingston, and Oleh Hornykiewicz, eds.) *Limbic Mechanisms*, Plenum Press, pp. 307–331.
4) MacLean, P. D. (1952). Some psychiatric implication of physiological studies on frontotemporal portion of limbic system (visceral brain), *Electroencephalogr. Clin. Neurophysiol.* 4, pp. 407–418.
5) Nauta, W. J. H. (1958). Hippocampal projections and related neural pathways to the midbrain in the cat, *Brain*, 81, pp. 319–340.
6) Nauta, W. J. H., and Domesick, V. (1982). Neural associations of the limbic system, in *The Neural Basis of Behavior* (A. I. Beckman, ed.). Spectrum Publications.
7) Nauta, W. J. H., and Domesick, V. B. (1981). Ramifications of the limbic system, in *Psychiatry and the Biology of the Human Brain: A Symposium Dedicated to Seymour S. Kety* (S. Matthysse, ed.). Elsevier, New York., pp. 165 188.
8) Meyer, A. (1971). *Historical Aspects of Cerebral Anatomy*, Oxford University Press.
9) Moruzzi, G., and Magoun, H. W. (1949). Brain stem reticular formation and activation of the EEG. EEG and Clin. Neurophysiol., 1, pp. 455.
10) Olds, J., and Milner, P. (1954). Positive reinforcement produced by electrical stimulation of septal areas and other regions of the rat brain *J. Comp Physiol. Psychol.* 47, 419–427.
11) Nauta, W. J. H., and Kuypers, G. J. M. (1958). Some ascending pathways in the brain stem reticular formation, in Reticular Formation of the Brain (H. H. Jaspers et al. eds.). Little, Brown and Co., pp. 3–30.
12) Guillery, R. W. (1957). Degeneration in the hypothalamic connexions of the albino rat. J. Anat. 91, 91–115.
13) Nauta, W. J. H. (1956). An experimental study of the fornix system in the rat. *J. Comp Neurol.* 104, pp. 247–271.
14) Brain, R. (1958). The physiological basis of consciousness: A critical review. Brain 81, 426–455.
15) Luria, A. R. (1973). The Working Brain: An Introduction to Neuropsychology, Basic Books.
16) Nieuwenhuys, R., Geeraedts, L. M. G., and Veening, J. G. (1982). The medial forebrain bundle of the rat. I. General introduction, J. of Comp. Neurol., pp. 49–81.
17) Powell, T. P. S. (1972). Sensory convergence in the cerebral cortex, in Connections of the frontal lobe with the limbic system, in Surgical Approaches in Psychiatry (L. V. Laitinen, and K. E. Livingston, eds.). University Park Press, Baltimore, pp. 266–281 *Proceedings of the Third International Congress of Psychosurgery*, University Park Press Baltimore.

18) Heimer, L. (1970). Selective silver-impregnation of degenerating axons and their synaptic endings in nonmammalian species, in *Contemporary Research Methods in Neuroanatomy* (W. J. H. Nauta, and S. O. E. Ebbesson, eds.). Springer Verlaag.
19) MacLean, P. (1990). *The Triune Brain in Evolution*, Plenum Press.
20) Nauta, W. J. H., and Feirtag, M. (1986). Fundamental Neuroanatomy, W. H. Freeman.
21) Butler, A., and Hodos, W. (1996). *Comparative Vertebrate Neuroanatomy*, Wiley-Liss.
22) Holstege, G. (1991). Descending motor pathways and the spinal motor system: Limbic and non-limbic components: in, *Progress in Brain Research* 87,. Elsevier Science Publications.
23) Lindvall, O. (1974). Mesencephalic dopamine neurons projecting to neocortex, in *Brain Research* 81, Elsevier Science Publications, pp. 325–331.
24) Heath, R. G. (1996). Exploring the Mind Brain Relationship, Moran Printing, Baton Rouge, Louisiana.
25) Corsi, P. (1991). The Enchanted Loom: Chapters in the History of Neuroscience, Oxford University Press.
26) Carlsson, A. (1987). Monoamines in the CNS: A historical perspective, in: *Psychopharmacology The Third Generation of Progress*, pp. 39–48.
27) Dahlstrom, A., and Fuxe, K. (1964). Evidence for the existence of monoamine containing neurons in the central nervous system. 1: Demonstration of monoamines in the cell bodies of brain stem neurones, *Acta physiol. scand.* 62, Suppl. 232, pp. 1–55.
28) Gloor, P. (1997). Temporal Lobe and Limbic System (1997) Oxford.
29) Anden, N. E., Dahlstrom, A., Fuxe, K., Larsson, K., Olson, L., and Ungerstedt, U. (1966). Ascending monoamine neurons to the telencephalon and diencephalon. *Acta Physiol. Scand.* 67, pp. 313–326.
30) Ungerstedt, U. (1971). Stereotaxic mapping of the monoamine pathways in the rat brain. Acta Physiol. Scand., 197, Supple. 367, pp. 1–48.
31) Nauta, W. J. H., and Domesick, V. B. (1978). Crossroads of Limbic and Striatal Circuitry: Hypothalamo-Nigral Connections in (K. E. Livingston, and O. Hornykiewicz, eds.). Limbic Mechanisms, Plenum Press, pp. 75–93.
32) Fuxe, K., Hokfelt, T., Johanssom, O., Jonsson, G., Lidbraink, P., and Ljungdahl, A. (1974). The origin of the dopamine nerve terminals in limbic and frontal cortex. Evidence for meso-cortical dopamine neurons, Brain Res., (82), pp. 349–355.
33) Nauta, W. J. H. (1971). The problem of the frontal lobe: A reinterpretation, *J. Psychiatric Research*, 8, pp. 167–187.
34) Kuhlenbeck, H. (1973). The Central Nervous System of Vertebrates, Vol. 5, Part 1, S. Karger.
35) Damasio, A. R. (1994). Descartes' Error: Emotion, Reason, and the Human Brain, Grosset/Putnam.
36) Barker, F. G. (1995). Phineas among the phrenologists: the American crowbar case and nineteenth-century theories of cerebral localization, *J. Neurosurg.*, 82, 672–682.
37) Macmillan, M. (1996). ISHN Annual Meeting abstracts http:/bri.medsch.ucla.edu/NHA/abs1996.html.
38) Star, S. L. (1990). What difference does it make where the mind is? Some questions for the history of neuropsychiatry, Journal of Neuropsychiatry and Clinical Neurosciences, V. 2 #4, 436–443. American Psychiatric Press.
39) Ward, C. D. (1990). Neuropsychiatry and the modularity of the mind. Journal of Neuropsychiatry and Clinical Neurosciences, V. 2 #4, 443–449. American Psychiatric Press.
40) Changeux, J. P. (1985). Neuronal Man. The Biology of Mind. Pantheon Books, New York.
41) Kinnier Wilson, S. A. (1940). Neurology, London Arnold.

CHAPTER FIVE

HEIMER'S LIMBIC STRICATUM

"... the objections could be raised that the projections from the limbic telencephalon are now known to involve not only the hypothalamus but also the so-called nucleus accumbens, a structure which by histological and histochemical criteria is unmistakably part of the striatum. This objection, of course, could be countered by the argument that the central nervous system is replete with examples of neural structures that from part of one 'system' when considered from one particular point of view, but just as plainly appear to be part of another 'system' when they are viewed from another perceptual angle, such ambiguities seem distressing only because of our tendency to expect a degree of separateness of neural mechanisms that appears to be an exception rather than the rule in the more highly developed central nervous system."

1) NAUTA, pp. 176–177

"... it would seem wise not to over emphasize the dichotomy between 'limbic' and extrapyramidal' mechanisms."

2) HEIMER, p. 191

SUBCORTICAL PROJECTIONS OF THE ALLOCORTEX: ANATOMICAL IMPLICATIONS FOR THE LIMBIC SYSTEM CONCEPT

"Originally, my interest was in psychosomatic medicine. The thinking, then, was that everything of this sort went on in the hypothalamus."

3) PAUL MACLEAN, p. 324

The hypothalamus as the common projection target for certain tele- nd diencephalic structures figured quite prominently as an integrating heme in the development of the limbic brain concept (3,4,5,6). Although uccessive constructs incorporated a larger collection of cortical and subcortical components (6,7,8) even in these constructs, admission of a neural component into the limbic organization was still dependent on the linkage of the candidate neural locus to a well defined subcortical neural con-tinuum centered on the hypothalamus (see epigraphs to chapter 4). This obeisance to hypothalamic centrality in limbic brain theory prompted one authority to criticize limbic brain thinking throughout the first three quarters of this century as Ptolemaic (9). And as Heimer pointed out such orthodoxy served as a constraining influence: the neocortex would perforce be related to the basal ganglia or extrapyramidal motor system through the corticostriato-pallidal pathway, while so called limbic structures were characterized foremost by their relation to the hypothalamus (5, p. 1).

In 1975, Lennart Heimer and Richard Wilson (2) published a paper which helped usher in a major shift in our conceptualization of limbic organization. Their paper demonstrated that contrary to the traditional anatomic wisdom, the allocortical centers (they investigated both the olfactory/piriform cortex, and the hippocampus) did *NOT* innervate the hypothalamus exclusively. These authors observed that a significant part of the neural afflux emanating from these structures engaged most prominently *NON* hypothalamic areas including the nucleus accumbens (a major recipient for hippocampal projections) and the medium-celled part of the olfactory tubercle (a major recipient of fibers from the olfactory cortex).

The results of Heimer and Wilson's findings not only helped to change traditional assumptions regarding limbic anatomy but also helped change our perception of how the basal forebrain is organized. In the following two subsections we will overview their findings.

Allocortical to Ventral Striatal Projections

Charting the results of heat lesions in both the piriform cortex, also the hippocampus, terminal degeneration was noted in the following structures the olfactory tubercle; ventral districts of the striatum; in cell bridges connecting these two structures,[1] also the mediodorsal nucleus of the thalamus Notably, the authors found a relative scarcity of degenerating nerve terminals in the lateral hypothalamus.

[1] Hippocampal projections to the ventral striatum had previously been described as early as 1943 by Fox (10). However, these earlier reports of such projections did not, to paraphrase Heimer, become part of textbook anatomy to the same extent as the hippocampal projections to the septum or the hypothalamus, (2, p. 179). Heimer attributed the relative obscurity of these earlier findings to their non-interpretability in terms of the then accepted principles of limbic system connections.

Since the olfactory turbercle and ventral striatum were identified by Heimer and Wilson as major projection fields from core limbic loci, it became an issue of some immediacy for the authors to both review and help refine our understanding of these two structures. Heimer and Wilson's reeaxamination led to a "realignment" of these structures with the brain's striatal organization, as opposed to septal or olfactory domains (2,5): indeed Heimer and Wilson held that the nucleus accumbens/ventral striatal continuum and the olfactory tubercle formed a part of a striatal macrostructure which extended all the way from the dorsal striatum to base of the brain. The arguments the author's put forth to support a striatal affiliation for the ventral accumbens and tubercle centered on the following points.

- *Tissue Continuity*: The olfactory tubercle, nucleus accumbens\ventral striatum and dorsal striatum are all continuous—indeed, the nucleus accumbens is an aspect of the ventral striatum. However, in macrosmatic animals the interposition of the deep olfactory radiation (the deep olfactory tract is represented in MacLean's diagram as the intermediate forebrain bundle, see figure 4.2 from chapter 4 of this volume) results in an 'artifactual' separation of the tubercle from the accumbens (2,14). The continuity between the tubercle and the accumbens, however, is still revealed by cell bridges linking the two structures across the deep olfactory radiation[4].

[2] There is no clear consensus on which part of the primate brain should be identified as the olfactory tubercle: it is generally considered to be a 'vestigial' structured imbedded in the anterior perforated substance of the primate. Notably, the anterior perforated substance is the port of entry of the penetrant vessels distributing from the circle of Willis, the brains vascular hilus. The olfactory tract fibers split into the medial and lateral olfactory stria (tracts) in the vicinity of the anterior perforated substance.

[3] According to Swanson and Cowan, Ariens Kappers, originally named the nucleus accumbens septi, it refers to the recumbent position of this nucleus against the septum (11). Nauta remarks that the recumbent position is more striking in the rat or cat, (12, p. 240). However, perhaps the term should be credited to Ziehen (13, p. 185). Importantly, as discussed in the text above, the nucleus accumbens septi is not functionally a part of the septal organization; its full name is misleading in this way. The nucleus accumbens and the surrounding ventral striatum are sometimes referred to as composing the fundus of the striatum.

[4] The rat is representative of forms that are macrosmatic but have limited neocortical investments. These forms do not evidence massive isocortical-diencephalic projections, and of consequence do not exhibit an internal capsule. The isocortical projection fibers are more dispersed—scattered throughout most of the striatum—and no separation is seen in the caudate. Separation of the caudate from the putamen is a feature of the higher primate brain. However, the macrosmatic rat does exhibit a rather robust olfacto-diencephalic projection system. This system is divides the ventral part of the striatum in a manner similar to the 'arbitrary' subdivision of the dorsal part of the striatum by the internal capsule in primates, for a discussion of this issue see (14).

- *Cell Architecture*: The nucleus accumbens, the medium-celled parts of the olfactory tubercle, and the cell bridges between them all evidenced a cell architecture characteristic of the caudate-putamen.
- *VTA-Dopaminergic Afferentiation*: The accumbens, tubercle and ventral striatal region were also united by the fact that they not only formed a continuous projection field for the allocortical fibers, but also for dopaminergic projection fibers from the ventral tegmental area.

Further support for a striatal alliance would come from experiments designed to trace the pathways from the accumbens, and tubercle: i.e., trace the exit lines from these structures. Perhaps these newly established allocortical to ventral striatal connections represented the first link in a cortical to striatal pathway that was similar to the well established dorsally positioned neocortical to striatal to pallidal pathways (see figure 5.3). If so, then the exit lines from the ventral striatum (similar to the more dorsally positioned pathway) should obtain the globus pallidus (figure 5.3). The authors set out to investigate this issue.

Ventral Striatal to Ventral Pallidal Projections (The 1st description of the v. pallidum)

Following heat lesioning to the olfactory tubercle and to multiple areas of the accumbens, Heimer and Wilson (employing silver methodology, also electron microscopy) reported terminal degeneration occurring ventral to and rostral to the anterior commissure. At the time Heimer and Wilson wrote their paper this area of the basal forebrain was so poorly understood, it was referred to as the unnamed substance (substantia innominata), see figure 5.1B. The terminal degeneration pattern the author's identified suggested that this aspect of the substantia innominata took part in the pallidal structure of the brain, and the author's dubbed this subcommissural district the ventral pallidum (2). Specifically, the ultrastructural synaptic anatomy in the ventral system appeared very similar to the dorsal system. Additional support for the pallidal nature of the subcommissural innominata would follow. This included the fact that the ventral and dorsal pallidum were shown to be continuous (see figure 5.2), and in receipt of substantial projection from the subthalamic nucleus; both areas shared a high iron content and both areas demonstrated strong enkephalin-like, and glutamic acid decarboxylase-like immunoreactivity.[5]

---

[5] It should be noted that in the same year Heimer and Wilson published their subcortical projections paper, Swanson and Cowan, using tritiated amino acids, also demonstrated accumbens to ventromedial pallidal pathways (11).

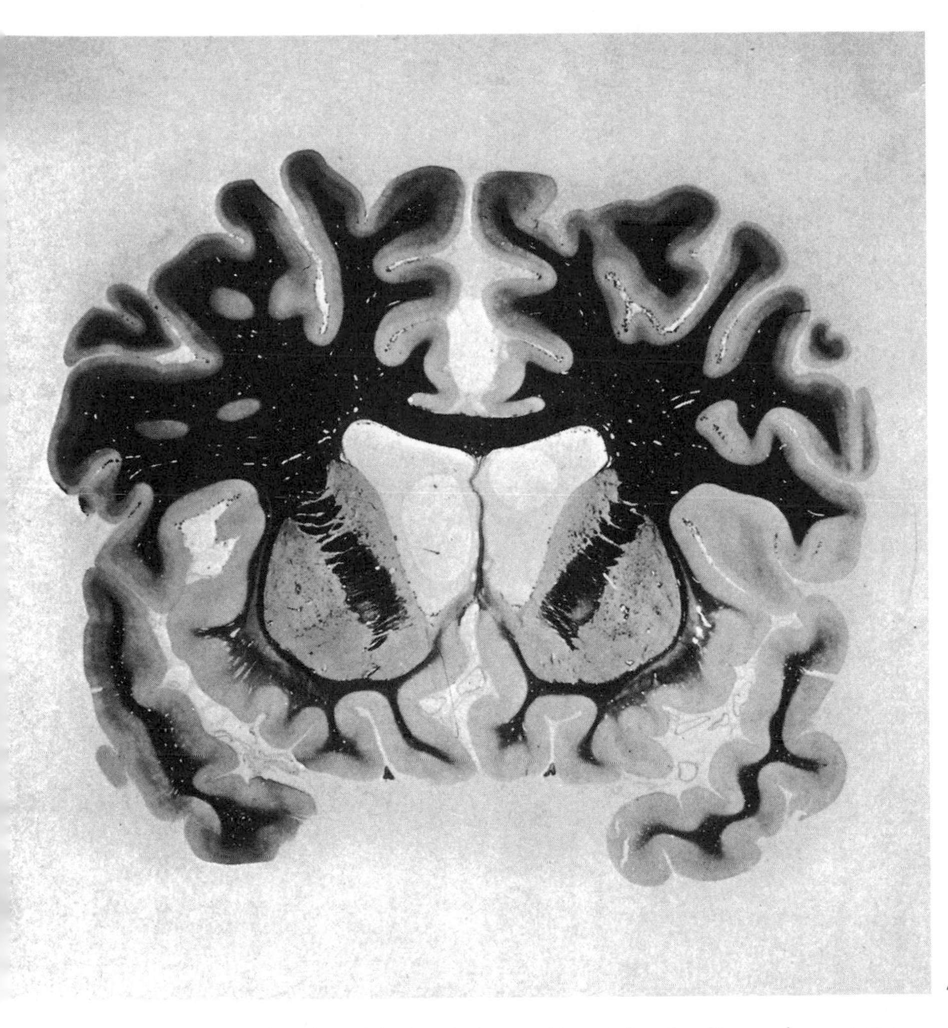

A)

Figure 5.1. A) A coronal section of the hemisphere, a bit rostral to the chiasma demonstrates the nucleus accumbens septi. Note the accumbens is located at the rostral inferior confluence of the caudate and putamen. B) A (diagramatic) coronal section at the level of the chiasma, also the crossing of the anterior commissure. The area just ventral to the anterior commissure is the *subcommissural* substantia innominata. The neighboring more posterior aspect of the substantia innominata, (it would be visible in a yet more posterior section), named for its position under the lenticular nucleus, is the *sublenticular* substantia innominata. In their paper, Heimer and Wilson, provided evidence that the subcommissural substantia innominata, should be understood as a ventral continuation of pallidal parenchyma. They named it the ventral pallidum. Notably, striatal parenchyma (ventral striatum) accompanied the newly described ventral pallidal district below the anterior commissure, all the way to the base of the hemisphere. Earlier generations of anatomists had explored the neural terrain below the anterior commissure, however, a convincing description of its anatomy always remained beyond reach. Heimer and Wilson set out to explore this same enigmatic neural landscape, the results of their efforts, in their own words: "The nucleus accumbens, olfactory tubercle, and rostral part of the substantia innominata, previously disparate parts of a murky "limbic forebrain," came into sharper focus as striatal and pallidal zones . . . ". Reproduced with permission (12).

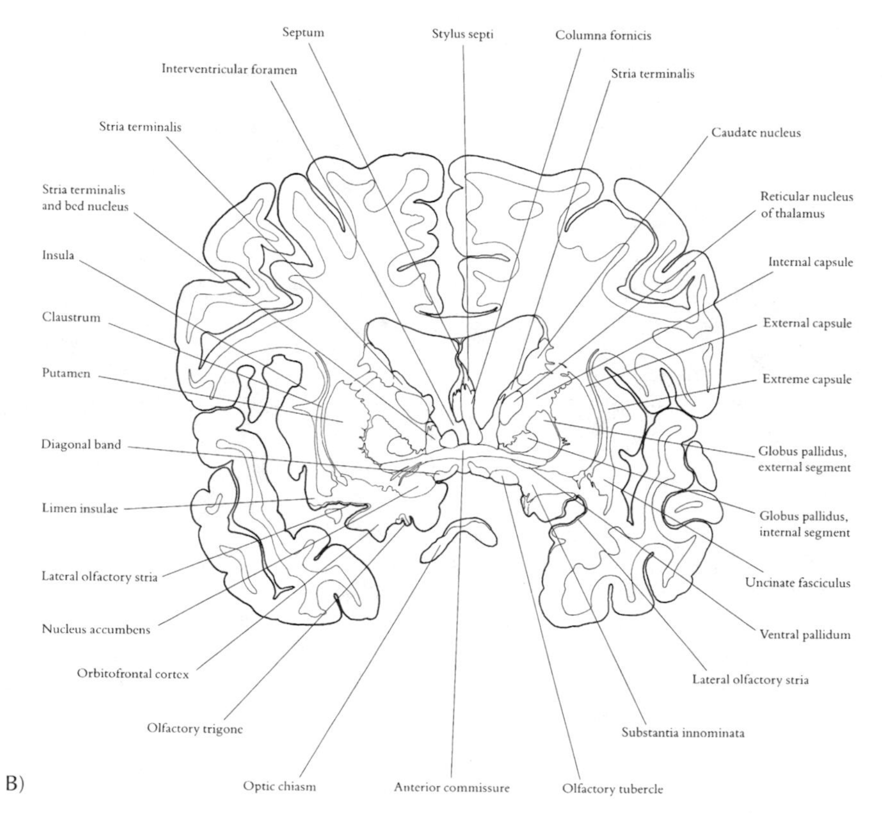

Figure 5.1. *Continued.*

Similarities Between The Subcortical Projection Fields of the Allocortex and Neocortex

It should be understood that a principal objective of Heimer and Wilson's paper was to explore subcortical projections of the allocortex, and one of the principal findings of their paper was to provide evidence for the presence of an allocortical parallel to the known neocortico-striato-pallidal pathway,[6] see figure 5.3. The change their paper brought to limbic brain

[6] As noted in the conclusion to their paper, their findings also served to emphasize that the basal ganglia might be better understood as a large-scale continuous cortical reception field, extending from the caudate to the floor of the telencephalon. The author's comment on the implication for such a perspective on basal ganglia function. "*On the other hand, the well-known influence of the basal ganglia on movements may reveal only the most easily visible of*

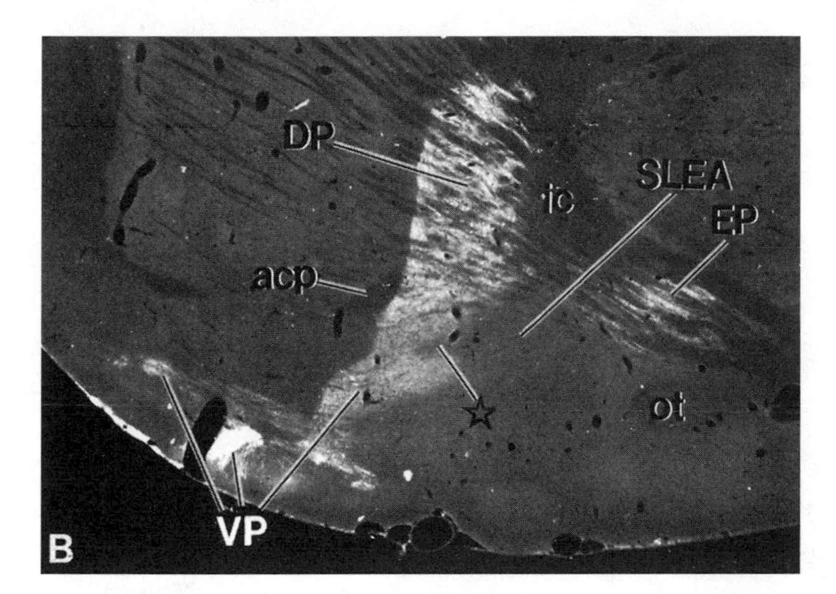

Figure 5.2. The dorsoventral extent of the rat pallidal complex is evident in this sagittal plane of the rat: note the continuity of the dorsal pallidum, (DP), with the ventral pallidum (VP). Also note how both the ventral pallidum and the ventral extension of the striatum are both contiguous and accompany each other to the ventral surface of the brain in the region of the olfactory tubercle. It should also be noted that the dorsal and ventral pallidal districts are divisions of the external pallidal segment. In the brains of certain mammalian orders (rodents and carnivores) the term globus pallidus refers ***Only to the external*** pallidal segment. In these forms, the internal pallidal segment, imbedded in the peduncular white, is referred to as the entopeduncular nucleus. Connections of the ventral striatum to the internal pallidal segment would also be discovered in follow up studies. An appreciation that the ventral striatum sent projections to both the ventral pallidum ***Also*** the internal pallidal segment meant that the ventral striato pallidal circuitry engaged all the major circuit elements of the pallidal system. Figure from (15).

theory is best understood as an aspect of this larger effort. Two decades after his subcortical projections paper, Heimer would summarize a principal achievement of his earlier publication.

> "The initial observation on the existence of the ventral striatopallidal system subserving allocortical and periallocortical areas simplified the analysis of the basal forebrain. The nucleus accumbens, olfactory tubercle,

---

its functions; perhaps the basal ganglia should be thought of not merely as components of the extrapyramidal system but rather as cell groups associated either with a type of neuron common to all cortices or with any group of cells in a complex cortical arrangement" (2, p. 190).

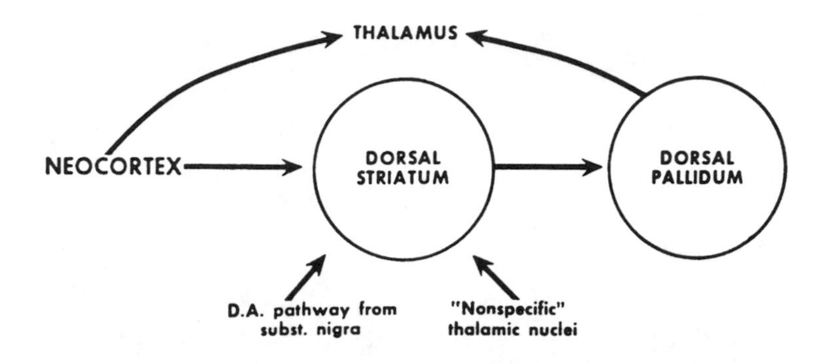

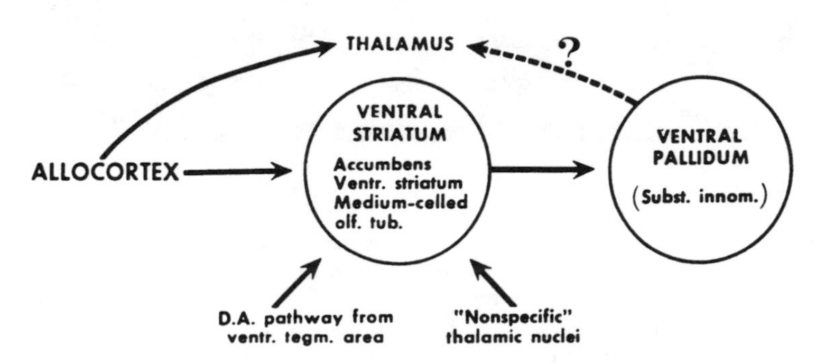

Figure 5.3. Heimer's schematic to illustrate the similarities between the cortico-subcortical relations of the neocortex and allocortex. "The similarities between neocortical and allocortical projections are striking, to say the least . . ." (2, p. 189). In their original paper, connections of the ventrostriatal system beyond the ventral pallidum were not pursued. Relying on the parallel between dorsal and ventral pathways, the authors suggested pallidal thalamic pathways (indicated by dotted arrow, with?) would be found. Anatomic theory had predictive value—within ten years of this paper, such connections were demonstrated (2).

and rostral part of the substantia innominata, previously disparate parts of a murky "limbic forebrain," came into sharper focus as striatal and pallidal zones, with predictable anatomical relations and transmitters based on our understanding of the regular features of the dorsal striatum."

(15, pp. 3–4)

"The ventral striatopallidal system, in general, provides neuronal connections for allocortex and for prefrontal and temporal association cortices, that are similar or complementary to those provided for neocortex by the caudate-putamen and globus pallidus. With the incorporation of the ventral striatum into the large striatal complex, it may be argued that all areas of cortex are served by subcortical target neurons."

(15, p. 5)

In their subcortical projections paper, the authors said little regarding possible behavioral correlation for the newly established limbic cortical to striatal connections.[7] However they did comment that their findings might provide a new perspective from which to approach physiologic studies of the basal forebrain. In the following sections, we examine the early development of correlative theory instigated in part by this new perspective on basal forebrain anatomy.

LIMBIC STRIATAL INTEGRATION—A MODEL FOR THE FUNCTIONAL INTERFACE BETWEEN LIMBIC AND MOTOR SYSTEMS

Emotion: [French émotion, from Old French esmovoir, to excite: Latin ex-, ex- + Latin movêre, to move)[8]]

Mood and Movement: Twin Galaxies of the Brain (the Inner Universe)

(17)

. . . it is the objective of this article to consider response initiation by limbic integrative processes, and in particular to consider a tentative model of the functional interface between limbic and motor systems."

18) G. MOGENSON and C. YIM, p. 76. 1980

[7] The decision of Heimer and Wilson to employ the term ventral striatum, and avoid the term limbic striatum (a term more likely to implicate behavioral correlation) serves to underscore the author's decision to side step behavioral issues in their subcortical projections paper. Notably, at the time there were compelling reasons to consider correlative (behavioral) issues. For example, the term limbic striatum had been employed as early as 1973 with particular reference to neurobehavioral correlates (13). Matthysse's article became a citation classic (*Current Contents, LS, V. 32, #34, Aug. 21, 1989*). The author's commentary in this citation classic provides a rich vein for the reader interested in early theory on brain behavior relations employing the term limbic striatum. Steven Matthysse would suggest a role for neuroleptic drugs at the nucleus accumbens in a series of articles dating to 1972 (16). Additionally, amygdala to striatal connections were well characterized by 1972, a finding cited by Heimer and Wilson in their paper (2).

[8] *The American Heritage® Dictionary of the English Language, Third Edition* Houghton Mifflin. Regarding etymological issues pertinent to mood and movement, it should also be noted (and it is often under appreciated) that that the term affect, a principal component of the psychiatric status examination, is a transitive verb, and refers specifically to the effect of the patient's psychomotor profile, degree of animation etc., on the observer. As Nauta once remarked. The only proof of motivation is movement.

> Conceptualization of the limbic and motor systems has permitted great advances in understanding the anatomical and functional connectivity within each system. However, our knowledge of how these two systems integrate to permit a smooth transition from emotional and cognitive awareness to adaptive behavioral responses is relatively underdeveloped. While it is impossible to assign a beginning to any field of study, momentum distinctly increased following Mogenson et al.'s classic formulation in 1980 entitled, "From Motivation to Action: Functional Interface Between the Limbic System and Motor System.
>
> (19, p. 238)

In 1980, Gordon Mogenson (1931–1991) co-authored an article, which tried to address and suggest a resolution for a major problem Mogenson felt was faced by behavioral neuroscience (18). This problem, according to Mogenson had to do with the lack of an anatomical mechanism to translate or transmit the motivational processes conducted by limbic forebrain structures and hypothalamus into the actual motor response. But as to the issue of limbic mediated motoric process, Mogenson found the literature wanting (18, p. 70). Mogenson attributed this to lack of an available anatomical mechanism that would serve in the limbic motor translation. It was Heimer and Wilson's proposal of allocortical to ventral striatopallidal connections that Mogenson (and others) felt would bridge this gap (20, p. 269).

Before we discuss the model of limbic motor integration put forth by Mogenson, it is helpful to briefly examine in simplified terms his views on motivation or initiation of behavior. Mogenson accepted a role for the cerebral cortex in response initiation—employing Kornoski's terminology movements and actions originating from the cerebral cortex were all considered to have come from the cognitive brain, (18, p. 78). As to the anatomical mechanism for activation arising in associational cortex (cognitive brain) obtaining striatal tissues, Mogenson referred to the then new findings of Kemp and Powell establishing corticostriatal to ventral thalamic circuitry (21).

However, Mogenson's more immediate focus was not on the associational or motor cortex initiating activity but more in the initiation of behaviors (such as innate drives related to biological adaptation and survival) by limbic loci—in the terminology of Kornoski, the limbic area was called the emotive brain. Mogenson's specific interest in response mediation by limbic integrative centers are summarized 1 below. As to the anatomical mechanism for activation arising in the core limbic-hypothalamic axis obtaining striatal tissues, Mogenson referred to Heimer and Wilson's subcortical projections proposal.[9]

The Basal Forebrain in Limbic Motor Integration

> "Locomotor and oral motor responses are of special interest in this article because they are fundamental components of food-seeking and ingestive responses, vocalization, escape from predators and survival. Limbic structures appear to have access to these pattern generators suggesting that the limbic system may have a direct role in the initiation of motor responses."
>
> (18, p. 77)

Mogenson exhibited a particular interest in basal forebrain mechanisms of survival and adaptation. We provide a partial list of Mogenson's thoughts in this area.

- *Primacy of Motor Activity*: Mogenson accepted motor activity as a dominant and primary component of all activity, (22, p. 194). The primacy of locomotor activity made it imperative for him to dissect out those component elements (circuit elements) elaborating locomotor activity.
- *Comparative Studies*: The basal forebrain and upper brain stem had long been suspected (even in the absence of direct experimental support) as a principal site of integration of visceral and somatic input (solitarius, dorsal vagal, trigeminothalamic system), and where this information was acted upon (i.e., the cranial motor nerves serving mastication, facial muscles effecting output). Mogenson specifically referenced Herrick's comparative investigations conducted in the early part of this century, that pointed to the possible role (in both rodents and mammals) of accumbens to pallidal connections in locomotor activity and facial reflexes involved in feeding (18, p. 85).
- *VTA to Accumbens Connections as a Critical Element in Limbic Mediated Responses*: Mogenson canvassed the neurophysiologic, behavioral, and anatomic literature all of which pointed to the role of neural loci in the basal forebrain, and midbrain, and their interconnections in the mediation of attack, ingestive and defensive behaviors. His literature review included (but far exceeded) the material that we have introduced in these chapters (i.e., Heimer and Wilson's paper, Nauta's limbic midbrain circuit, Bard's transection work) also his own work in

[9] Mogenson was not the first to incorporate newer anatomic concepts into a theory of limbic motor integration: similar proposals incorporating the ventral striatum and ventral pallidum were put forth by Graybiel, also McGeer, the work of both of these investigators is acknowledged by Mogenson in his paper.

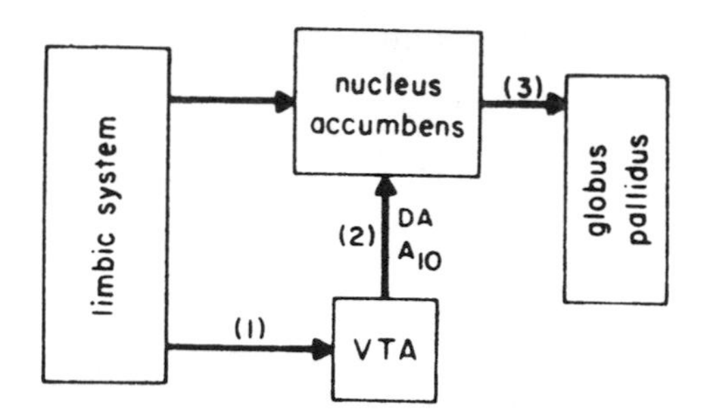

Figure 5.4. The Wiring Diagram. As underscored by Mogenson, the amygdala, septal, area, hippocampus, and other limbic forebrain structures are connected with the hypothalamus and midbrain by such pathways as the stria terminalis, fornix, and MFB. Mogenson's experimental efforts helped lead to an appreciation of how these circuitry elements and not just the nuclei or neural loci played an important role in the translation of motivation into action (18).

electrical stimulation (stimulation of the preoptic area affected changes at VTA) and microiontophoresis (examination of the effects of iontophoretically applied dopamine in the VTA, also accumbens)[10] (18). Mogenson's thoughts regarding the convergence of limbic and striatal transmitters is underscored in the following passage. "The hippocampus represents a major limbic structure that has been shown to be functionally involved in the initiation of locomotor movements in exploration and in the processing of spatial memory. Excitatory glutaminergic hippocampal outputs projected to the accumbens where dopaminergic inputs from the VTA also converge. Conceivably, hippocampal signals coding for spatial orientation, and VTA signals coding for reward significance, can be integrated into motor behavioral output." (p. 194, 22).

Wiring Diagram—The Model

In the grand tradition of the diagram makers, a tradition in neurologic science dating to the epoque of localization in neurology, Mogenson presented a tentative model of the functional interface between limbic and motor systems in the basal forebrain, see figure 5.4. For a discussion of limbic motor circuitry see (15,19,20,22,23).

---

[10] Interestingly, Mogenson doesn't appear to cite contemporary efforts of such workers as Olds, and Routtenberg, which established the importance of the MFB in cortical and subcortical brain mediated reward paradigms.

CONCLUSION

We end this introduction to the limbic brain with a discussion of the limbic striatum. The aim throughout these chapters has not been to support an essentialist position in which to place the limbic brain, but to try to place the concept in an historical context; and examine basic principles of limbic brain anatomy. Located at the hemispheres hilum, the study of the limbic brain can serve as a platform to examine basic principles of hemispheric anatomy. Located at the center of a century and a half of debate regarding brain and consciousness, and relatedly, emotional function, the study of the limbic brain has evoked and should continue to provoke inquiry into the most basic principles of structure and function.

REFERENCES

1) Nauta, W. J., and Domesick, V. B. (1982). Neural associations of the limbic system, in *The Neural Basis of Behavior* (A. L. Beckman, ed.), SP Medical & Scientific Books, New York, pp. 175–206.
2) Heimer, L., and Wilson, R. D. (1975). The Subcortical Projections of the Allocortex: Similarities in the Neural Associations of the Hippocampus, the Piriform Cortex, and the Neocortex, in *Golgi Centennial Symposium. Proceedings* (M. Santini, ed.), Raven Press, New York, pp. 177–193.
3) MacLean, P. D. (1986). Epilogue: Reflections on James Wenceslas Papez, according to four of his colleagues, in *The Limbic System* (B. K. Doane, and K. E. Livingston, eds.), Raven Press, New York, pp. 317–334.
4) Papez, J. W. (1937). A proposed mechanism of emotion, *Arch. Neurol. Psychiatry*, 38, pp. 725–743.
5) Heimer, L., and Alheid, G. F. (1991). Piecing together the puzzle of basal forebrain anatomy, in *The Basal Forebrain* (T. C. Napier, P. W. Kalivas, I., & P., Hanin, eds.), Plenum Press, New York, pp. 1–41.
6) MacLean, P. D. (1949). Psychosomatic disease and the "visceral brain." Recent developments bearing on the Papez theory of emotion. *Psychosomatic. Med.* 11, pp. 338–353.
7) MacLean, P. D. (1952). Some psychiatric implication of physiological studies on frontotemporal portion of limbic system (visceral brain), *Electroencephalogr. Clin. Neurophysiol.* 4, pp. 407–418.
8) Nauta, W. J. H. (1958). Hippocampal projections and related neural pathways to the mid-brain in the cat, *Brain*, 81: 319–340
9) Swerdlow, N. R., Braff, D. L., Barak, Caine, S. B., and Geyer, M. A. (1993). Limbico cortico-striato-pontine substrates of sensorimotor gating in animal models and psychiatric disorders, in *Limbic Motor Circuits and Neuropsychiatry* (P. W. Kalivas, and C. Barnes, eds.), CRC Press: 311–349.
10) Fox, C. A. (1943). The stria terminalis, longitudinal association bundle and precommissural fornix in the cat. *J. Comp. Neurol.* 79: 277–295.
11) Swanson, L. W., and Cowan, W. M. (1975). A note on the connections and development of the nucleus accumbens, *Brain Research*, 92: 324–330.

12) Nauta, W. J. H. (1986). Fundamental Neuroanatomy, W. H. Freeman.
13) Stevens, J. (1973). An Anatomy of Schizophrenia? Archives of General Psychiatry, V. 29: 177–189.
14) Heimer, L., Switzer, R. D., and Van Hoesen. G. W. (1982). Ventral striatum and ventral pallidum: Components of the motor system? *TINS*, 5, pp. 83–87.
15) Heimer, L., Alheid, G. F., and Zahm, D. S. (1993). Basal forebrain organization: An anatomical framework for motor aspects of drive and motivation, in *Limbic Motor Circuits and Neuropsychiatry* (P. W. Kalivas, and C. Barnes, eds.), CRC Press, pp. 1–43.
16) Matthysse, S. (1973). Antipsychotic drug actions: a clue to the neuropathology of schizophrenia? *Federation Proceedings* Vol. 32, #2, pp. 200–205.
17) McGeer, P. L. (1976). Mood and movement: twin galaxies of the inner universe. *Proc. Soc. Neurosci,*. Invited Lecture. Toronto Canada.
18) Mogenson, G. J., Jones, D. L., and Yim, C. Y. (1980). From motivation to action: Functional interface between the limbic system and the motor system, *Progress in Neurobiologyy*, pp. 69–97.
19) Kalivas, P. W., Churchill, L., and Klitenick, M. A. (1993). The circuitry mediating the translation of motivational stimuli into adaptive motor responses, in *Limbic Motor Circuits and Neuropsychiatry* (P. W. Kalivas, and C. Barnes, eds.), CRC Press, pp. 237–287.
20) Mogenson, G. J., and Yang, C. R., (1991). The contribution of basal forebrain to limbic-motor integration and the mediation of motivation to action, in *The Basal Forebrain* (T.C. napier, P. W. Kalivas, I., & P., Hanin, eds.), Plenum Press, New York, p. 267–290.
21) Kemp, J. M., and Powell, T. P. S. (1971). The connexions of the striatum and globus pallidus: synthesis and speculation. *Phil. Trans. R. Soc. Lond.* (B) 262, 441–457.
22) Mogenson, G. J., Brudzynski, S. F., Wu, M., and Yang, C. R. (1993). From motivation to action: A review of dopaminergic regulation of limbic → nucleus accumbens → ventral pallidum → pedunclulopontine nucleus circuitries involved in limbic-motor integration, in *Limbic Motor Circuits and Neuropsychiatry* (P. W. Kalivas, and C. Barnes, eds.), CRC Press, pp. 193–236.
23) Groenewegen, H. J., Berendse, H. W., Meredith, G.E., Haber, S. N., Voorn, P, Wolters, J. G., and Lohman, A. H. M. (1991). Functional Anatomy of the Ventral, Limbic System-Innervated Striatum, in The Mesolimbic Dopamine System: From Motivation to Action. John Wiley.

INDEX